Agriculture Issues and Policies

www.novapublishers.com

Agriculture Issues and Policies

Emerging Technologies to Combat Biotic Stress in Crop Plants and Food Security
Anirban Bhar, PhD (Editor)
Aryadeep Roychoudhury, PhD (Editor)
2023. ISBN: 979-8-89113-079-1 (Hardcover)
2023. ISBN: 979-8-89113-137-8 (eBook)

The Future of Cropping Systems
Julia E. Daniels (Editor)
2023. ISBN: 979-8-89113-011-1 (Softcover)
2023. ISBN: 979-8-89113-072-2 (eBook)

Pomegranate: For Horticulture Students and Farmers
Arkalgud Nanjundaiah Ganeshamurthy, PhD (Editor)
2023. ISBN: 979-8-88697-812-4 (eBook)

Research Advancements in Organic Farming
Jatindra Nath Bhakta, PhD (Editor)
Sukanta Rana, PhD (Editor)
2023. ISBN: 979-8-88697-519-2 (Hardcover)
2023. ISBN: 979-8-88697-580-2 (eBook)

Legumes: Nutritional Value, Health Benefits and Management
Phetole Mangena, PhD (Editor)
2023. ISBN: 979-8-88697-558-1 (Softcover)
2023. ISBN: 979-8-88697-583-3 (eBook)

More information about this series can be found at
https://novapublishers.com/product-category/series/agriculture-issues-and-policies/

Youssef Lee
Editor

Agricultural Policy

Strategies, Challenges and Global Implications

NOTICE TO THE READER

Library of Congress Cataloging-in-Publication Data

Names: Lee, Youssef, editor.
Title: Agricultural policy : strategies, challenges and global implications / Youssef Lee (editor)
Other titles: Agriculture issues and policies series
Description: New York : Nova Science Publishers, 2024 | Series: Agriculture issues and policies | Includes bibliographical references and index |
Identifiers: LCCN 2024017284 (print) | LCCN 2024017285 (ebook) | ISBN 9798891137202 (paperback) | ISBN 9798891137875 (adobe pdf)
Subjects: LCSH: Agriculture and state. | Rural development. | Agricultural development projects.
Classification: LCC HD1415 .A3183 2024 (print) | LCC HD1415 (ebook) | DDC 338.1/8--dc23/eng/20240514
LC record available at https://lccn.loc.gov/2024017284
LC ebook record available at https://lccn.loc.gov/2024017285

Published by Nova Science Publishers, Inc. † New York

Contents

Preface

This book consists of five chapters that focus on recent agricultural policy. Chapter one argues that adaptation and households' responses to changes in the environment are complicated by the fact that human actions do not occur in a social vacuum and are often mediated by a variety of historical and social factors, which are socially embedded. Chapter two explores the significance of wage employment to rural livelihoods and how the enduring formal unemployment crisis, witnessed since the 1990s, has affected peasant agriculture. Based on an application of Mancur Olson's collective action approach, chapter three draws attention to the role of Northern development NGOs in advancing agricultural protectionism. Using Malawi as a case study, chapter four contributes to the contemporary debates on the youth employment challenge, particularly those centered on the potential role of agriculture in solving youth unemployment, accelerating economic growth, and accelerating intra-continental trade. Chapter five provides a brief scrutiny of major issues in agricultural policy analysis aiming at contributing to informed policy debate in this subject area.

Chapter 1

Socio-Cultural Limits to Adaptation: Small-Farm Households, Agricultural Practices and Changing Natural Environment in North-Western Zimbabwe

Vusilizwe Thebe*

Department of Anthropology and Archaeology, University of Pretoria, Hatfield, Pretoria, Republic of South Africa

Abstract

Debates on climate change and variability often emphasise impacts on human systems, adaptation and adaptive strategies, and the significance of developing adaptive capacity of human societies for them to withstand shocks and stresses associated with climatic variability. Through evidence from a study of a small-farm community in a semi-arid former reserve in north-western Zimbabwe, this chapter challenges these assumptions, arguing that adaptation and households' responses to changes in the environment are complicated by the fact that human actions do not occur in a social vacuum, and are often mediated by a variety of historical and social factors, which are socially embedded. The chapter shows how farm households interpret and respond to changing weather patterns and declining soil fertility, and practice their farming, are informed by their past experiences of weather pattern, the farming system, and social systems of rituals, and fear of social sanctions. This highlights the limits to adaptation to climate change, and the implications for policy on climate change in these small-farm communities.

* Corresponding Author's E-mail: vusi.thebe@up.ac.za.

In: Agricultural Policy
Editor: Youssef Lee
ISBN: 979-8-89113-720-2

Keywords: climatic changes, human systems, small-farm households, socio-cultural dimensions of farming, Zimbabwe

Funding: NRF GRANT for 2021: Competitive Programme for Rated Researchers
Grant No. 129302

Conflicts of interest/Competing interests: The authors of the articles have no conflict of interest, whether with regard to the journal or publisher

Ethics approval: This data is derived from a broader study on 'Former Labour Reserves and Questions of Food Security'. The study was conducted under the University of Pretoria as part of staff research, and Ethical approval was provided by the University of Pretoria's Faculty of Humanities Ethical Committee. The author is a faculty member.

Introduction

Debates on climate change and variability often emphasise impacts on human systems, adaptation and adaptive strategies. Since climate is a primary determinant of agricultural productivity, changing climate conditions can potentially affect agriculture (Adams et al., 1998), as well as increase the vulnerability of farming communities. Agriculture has been identified as one of the most vulnerable sectors to climate change.

The main concerns have been about developing adaptive capacity of human societies to withstand shocks and stresses that are associated with varying weather conditions. Adaptation in this case is associated with "adjustments in individual groups and institutional behaviour in order to reduce society's vulnerability to climate" (Pielke, 1998:159).

To this end, numerous human inventions have been developed as mechanisms to assist farm households to adapt to changing climate conditions. Although the human species is the most adaptive to changes in its environment, and has been adapting for time immemorial, the phenomenon of climate change appears more complex and much bigger than any other change that the human species has experienced in the past.

While the focus on adaptation generates some optimism about the livelihoods and survival of small-farm households in the global south (the southern African region included), increased incidents of food insecurity and extended humanitarian programmes, continue to generate great pessimism. There is still scope for further exploration of how small farmers experience, respond or adapt their farming activities in the face of experience and culture.

Current emphases on adaptation may be misleading insofar as they presuppose that farmers can overcome their vulnerability to changing climate, by developing adaptive capacities. This chapter argues that, by making such assumptions, there is a risk of underestimating the power of the social environment in shaping responses to changes in one's environment. This chapter adds to the growing body of knowledge on small-farm households and adaptation to climate change, through the case study of a community of small farmers in semi-arid Lupane District, north-western Zimbabwe.

It particularly highlights the limits to adaptation to changing weather conditions, imposed by certain cultural dynamics including prior experiences and knowledge systems, and societal values. It argues that adaptation occurs and is mediated by these socio-cultural limits rather than anything outside, and in policy terms, understanding these cultural dimensions is key to understanding households' vulnerability.

Sections 2 and 3 of this chapter provide a conceptual argument and background. Section 4 provides the methodology, while the sections that follow engage with dynamics at the case study level, highlighting in particular the cultural dimensions of farming and households' responses to climatic changes.

Adaptation to Climate Change, Culture and Limits to Adaptation

Climate change, broadly understood, has become one of the major pre-occupations for stakeholders – the international community; developed and development country governments; civil society organisations; and scholars working in environment, development and agriculture – in the past 20 years or so. I use the term climate change here fully aware of its complexity, noting in particular its cross-disciplinary appeal. I draw here on Catherine Leyshon and Hilary Geoghegan's definition of climate change as:

> The circumstantial, suggestive, remembered and observed changes to weather and seasons that form the basis of an understanding of what is changing, if not why (Leyshon and Geoghegan, 2013:463).

Thus, I move the concept from the domain of the natural sciences, recognising in the process its diffusion into the social sciences. Clearly such definitions of climate change highlight the importance of also paying attention to the cultural dimensions of climate change (Brondizio and Moran, 2008), and how "climate change impacts…are given meaning through cultural interpretations of science and risk" (Adger et al., 2013:112). As Simon Batterbury (2008:62) noted:

> Climate variability and weather patterns are a permanent backdrop of, and often a central force in, social life. Climate changes, the longer-term shifts that take place in a climate regime, are central to the fate of cultures, ecosystems and regions.

The centrality of culture in how humans understand (the causes, meaning and manifestation) and respond to climate change remains key to policy (Adger et al., 2013). Certainly, human society of all kinds continuously strives to reduce vulnerability to changes, by building a combination of mitigation and adaptation mechanisms and strategies (Mulinge and Getu, 2013).

Indigenous knowledge systems may well be common wisdom at the community level, and have been a major resource for communities since time immemorial. Through such systems, communities in Africa, including southern Africa, have coped well with climatic events such as droughts, floods, etc., for many generations. However, systems get stretched "where changes may be significant and pervasive and incorporate elements of surprise through the occurrence of extreme events" (Thomas et al., 2007:302).

As Adger et al., (2013:113) noted: "Although local knowledge and practices can be effective for progressively adapting to climate change, they may have limited utility when cultures are confronted with rapid or nonlinear changes". Using the case of Pueblo Indian people, who were successful in adapting to drought through the use of various strategies, they noted how the effectiveness of such strategies became compromised as droughts became more prolonged and intense.

Cultural ecologists have alerted us to how culture mediates how humans identify risks, how they respond, and the means through which they implement adaptation, and as such, the importance of incorporating the perspective of the people into climate and environmental change debates. This, and what it takes to achieve it, is what led to the expansion of the field of anthropology in climate and environmental change (for example, Brown, 1999; Orlove, 2005).

The growing increase in anthropological interest, and emphasis on the local knowledge, are part of this discourse, and so are efforts by development agencies in other regions to incorporate social science research in their projects (Batterbury, 1996; Crate, 2011; Falk Moore, 2001).

One particular area in all this that matters to this analysis is the focus on the social and local dynamics through which human action is mediated, and, secondly, the fact that these factors, which include societal values, the knowledge that individuals construct and the existing relationships that exist between individuals and social institutions, pose limits to the individual to adapt (Adger et al., 2009).

As Adger et al., (2013) have argued, socio-cultural factors can act as enablers or major impediments to adaptation, particularly if there are tensions between adaptations introduced by outsiders and the desires of individuals and societies. Acceptance and subsequent adoption of a particular adaptation depends largely on whether such adaptation violates certain values or value systems.

These can be described as "ethics (how and what we value), knowledge (how and what we know), risk (how and what we perceive) and culture (how and why we live)" (Adger et al., 2009:338). These, according to O'Brien and Wolf (2010:233): "….serve as standards or criteria to guide not only action but also judgment, choice, attitude, evaluation, argument, exhortation, rationalisation, and, one might add, attribution of causality".

In considering whether an adaptation is worth adopting or not, an individual will be guided by assumptions and beliefs of his world. This can easily lead to a shift from adoption and implementation to indifference and at the extreme, resistance. One way of looking at the choice between adaptation and non-adaptation is to view individuals as rational beings that engage in calculations – calculations of risks imposed by climate change to values and other cultural expressions that are considered important in society (Adger et al., 2013).

If climatic changes exacerbate risks to culture, as they and I argue, then individuals in their different capacities will develop their adaptation responses based on their knowledge and experiences and perception of that risk, including whether the risk is worth the change or adaptation (Adger et al., 2009, 2013).

The Zimbabwean Context

Although recent climate trends in Zimbabwe suggest that the country is experiencing climate change, adaptation remains a major challenge for farmers that are still heavily dependent on "rain-fed agriculture and climate sensitive resources" (Brown et al., 2012:ii). Zimbabwe continues to experience rainfall variability and extreme weather events including droughts. Over the years, it has become increasingly difficult for rural farmers to construct livelihoods through agriculture since climate change have expanded arid environments with resultant shifts in the country's five agro-regions (Brown et al., 2012). Prolonged droughts have made the production of maize difficult because of its intolerance to drought (Andersson, 2007; Brown et al., 2012).

Successive droughts and high unemployment, combined with an increasing proportion of women headed households, have substantially compromised the ability of households to respond to shocks. Consequently, poor rural households, particularly those located in drier regions have plunged into a livelihood and humanitarian crisis as they had to shed assets including cattle. This strategy exacerbated their vulnerability as they eroded their resource base and compromised future food security (Mazzeo, 2011). Households in this category have fewer options for diversification in the context of a declining economy, often characterised by reduced employment opportunities (Bird and Shepherd, 2003; Mazzeo, 2011).

Although rural households have often recovered from shocks, the perennial nature of the droughts, particularly in the 2000s, have compromised such ability as assets were drawn during prior periods of stress (see Bird and Shepherd, 2003). Mazzeo (2011) reported incidents where farmers had to draw on cattle for cash to access food following the 2004/2005 drought, despite their strong dependence on the animals for future livelihoods.

In the context of increased vulnerability and reduced production, households cloaked the food aid and relief system, with the World Food Programme reporting a figure of 1.68 million people who received humanitarian aid in the form of subsidised agricultural inputs and food aid in 2010 (WFP, 2010).

Zimbabwe's total land area of 391, 000 square kilometres divides into five main agro ecological regions of different agricultural potential and rainfall patterns. Rainfall progressively declines from Natural Region I to V. Arid regions are by far the largest in the country, and with the reported expansion of arid environments and shifts in agro-regions, these have apparently increased (Brown et al., 2011). Although all regions experience variable climate, arid regions are the most affected, and are generally prone to droughts (Bird and Shepherd, 2003; Brown et al., 2011).

Precipitation varies widely across the country's regions and depending on location on the agro-ecological continuum, in some areas rains are too low for any meaningful production of popular crops under rain-fed conditions (Brown et al., 2011). Arid zones typically have a shorter rain season, and water presents by far the greatest constraint to crop production (FAO, 2005).

Although the country as a whole, like the rest of the region, experiences El Niño Southern Oscillation (ENSO) phenomenon every 10 to 20 years (Kandji et al., 2006), arid regions have fared badly. Weather conditions, however, can determine livelihood choices. For example, in parts of the arid belt that was established as labour reserves, agriculture is mostly part-time and practiced by women and the elderly, while men migrate to cities and mines to make money (see Thebe, 2018). In these regions, both part-time and full-time farmers produce maize, despite its high moisture requirements (Andersson, 2007).

In contrast to other regions in the country, conditions in the semi-arid belt are conducive for the production of drought tolerant crops and livestock husbandry, and these have long been encouraged as a solution against erratic rains. Yet, households have been known for their preference and persistence with the production of maize at the expense of alternative cereals that are not sensitive to water stress (Mavhura et al., 2015). Small-farm households produce for subsistence, and, traditionally have produced drought tolerant crops such as pearl millet, sorghum and melons, to cope with drought conditions. Entrepreneurial farmers would grow commercial crops, but even within this group, there were little differences.

Methods and Sampling

This paper draws on an extended study conducted between December 2016 and July 2018 on 30 small-farm households and their responses to changing natural conditions and the new demands on their farming practices in Gumede Communal Area in Lupane District, north-western Zimbabwe.[1] The study was part of a broader project focused on understanding dynamics of change in

[1] Gumede is part of the Menyezwa Ward.

former migrant labour societies in southern Africa. It also drew material relevant to the study from previous studies conducted in the area since 2005.

I became interested in households and how people relate to their environment following incidents of land self-provisioning in the area (see Thebe, 2017a). I also carried out short periods of ethnographic research on two villages at critical periods of the agricultural season (planting and harvesting) in the 18 months between December 2016 and July 2018. The techniques employed included community tours and non-participant observations at the village level, and random household visits, open-ended discussions and participant observations at the household level. The intention was to understand the social and natural environment: weather patterns and land, farming practices, social factors that inform farming decisions, and responses to changes in the natural environment.

The study also explored the wider context under which natural manifestations such as droughts were understood and interpreted and the interventions available to society as a whole, and how farm households' responses were mediated by the broader social context. The intention was to enable an analysis of social and cultural dimensions of farming, but also, it was oriented towards how these factors limit adaptation to changing natural conditions experienced in the recent past.

Gumede provided an ideal setting for studying farm households, environmental change and households' responses for mainly three reasons: First, it is semi-arid, and like the rest of the '*gusu*' frontier, has a short rain season; it frequently experiences erratic rains, dry spells and drought conditions.[2]

Second, the dominance of sandy soils (the '*gusu*' of Kalahari sands) and clayey loams, together with semi-aridic climatic conditions, means that the area is best suited for the production of *sorghum* and fodder crops, while there are limits to the production of maize without supplementary irrigation. Third, these climatic characteristics and frequent droughts have led to interventions by NGOs and other government agents which have promoted certain agricultural technologies including minimum tillage models.

The selection of households was based on a criteria-based purposive sampling technique. The main focus was those households that are located on the three different types of soils – sandy, clayey, and mixed – in the settlement area, households that had self-allocated themselves land, and those with fields

[2] The "dark *gusu*" that was explicitly described by Alexander et al., was one of the native reserves created by the colonial government for Ndebele natives.

only in the settlement area. Since the study focused on farming practices and responses to changing natural conditions, non-farming households and households without land were not considered for participation in the study.

Results and Discussions

Gumede and Its Natural Environment

Gumede Village, in southern Lupane District on the southern fringes of the former Shangani Reserves, occupies a strip of valley between the Gwayi River (to the south) and the A8 Highway to the north in the 'dark *gusu*' frontier (Alexander et al., 2000). In agro ecological terms the area, similar to the rest of the frontier, is located on Agro-ecological Region IV, which is agriculturally marginal, although dry-land farming still occurs during rainy seasons. With annual rainfall of between 450 and 650mm, crop failure is quite common. The area is characterised by erratic precipitation, with rainfall varying from year to year, and sometimes resulting in persistent droughts.

In good years, the area experiences annual rainfall of about 700mm, while in bad years annual rainfall may be lower than 450mm. As with all other areas in the region, the calendar divides into four major seasons of differing weather conditions: the rains begin around August, and the growing season starts in late October and extends into the month of May. The winter period, which begins in May and ends in July, is generally dry, and rains are not expected during this period.

While climatic uncertainty has always been a common feature, households have had harsher experiences with weather changes in the past two or so decades than in other years. In the past, rain patterns were predictable and clear, with good to moderate rains in November, December and February; long dry spells during the month of January; declining precipitation between March and May; and a drier month of June. Rains in November and December were important for early crops, while the February rains supported crops that were planted late.

The rain patterns also allowed farmers to stagger their crops, and provided mitigation against crop loss due to under-precipitation during any one of the rain phases. Since 2000, however, droughts have been perennial that it has been difficult for households to sustain from their own farming. There have been several years of less than optimal rain in the post-2000 period.

Over most of the '*gusu*' region, the *Kalahari*-type sandy soils are well represented, being significantly dissected by clayey and mixed loams on riverine and river valleys.[3] Gumede Village is located to the south of the Gwayi/Shangani watershed, and the strong water current flowing down the gradient towards the Gwayi River, causes massive damage to the land. The village occupies four common types of soils found in the '*gusu'* frontier region – the sandy soils, red clay, black clay and mixed soils – with the bulk of households occupying land on the sandy and mixed soils. Only a minority of households have fields on red clay soil, while the most fertile land – the black clay – lie outside the cropping areas (in the reserved arable zone, south of the settlements and former commercial ranch, south of the Gwayi River.

The former are fine, sandy loams, which are brownish in colour, and because of "poor nutrient availability and low nutrient and water retention capacity", they are generally poor for crop production (Andersson, 2007:683). These have low tolerance and resistance to erosion since they are easily eroded due to their poor retention and protection of organic matter, which is crucial in retention of water and nutrients. Over the years, these qualities have been enhanced through the application of organic manure, which remains a major source of organic matter in the region and country in general.

While primarily a "dark '*gusu'* forest" inhabited by a forest tribe (the *sili*) (Alexander et al., 2000), these parts of the former Shangani Reserves became home to Ndebele tribal groups, which moved into the '*gusu'* frontier following evictions from "white land" after WWII. The availability of uncontested land for crop production and cattle husbandry on the one hand and the road network and transport connecting the area to Bulawayo on the other, shaped a society that was largely semi-proletarian. In this part of the frontier, farming was centrally about subsistence and securing rural land.

Although ideally crop production or self-sufficiency is an aspiration for many rural households, particularly in agrarian societies, the reality in the frontier general, was that households experienced many challenges and disappointments on two fronts - soil fertility and climatic conditions. An agrarian path would likely leave households vulnerable to food insecurity. The agriculture season was characterised by adverse weather patterns including varying climate, erratic rainfall and droughts, which ultimately resulted in harvest failures. As a result, households found themselves occupying a blurry position that did not conform to traditional stereotypes of agrarian societies.

---

[3] The forest frontier where the Ndebele ethnic groups were resettled is dominated by sandy soil, which the local refer to as the "*gusu*".

The men who already held jobs in the urban centres and those who were entering adulthood spent most of the time at work and circulated between the urban and the rural, depending on their employment situation. The involvement of men in agriculture was also despised since agriculture was not remunerating. Among its consequences, the system contributed to a highly gendered division of labour, wherein men engaged in migrant labour, and women farmed the land with the help of children. Historical accounts reveal that households that depended entirely on their own farming were of lower social status than those that had supplementary income from wage labour.

In both set of households (farming and labour migrant), the plough was a significant tool in the production of crops, and failure to access a plough severely limited the area designated to crops. The farming system, particularly with respect to the use of the plough, has been linked to the history of the households in the former "white lands" (see Alexander et al., 2000). The use of the plough, oriented households towards agricultural extensification, while ownership of cattle also meant that organic manure became a central component of the agriculture system. However, it was also recommended by the country's agricultural extension service as key to soil fertility.

Although a majority of households derived the bulk of support from the wage, all households had cropping fields, and lay claims to large tracts of land (see Thebe 2012). In addition, all households had access to the commons. These commons were also used for grazing, although the Village Headmen could still allocate this land to new households. All the land that was initially allocated to households belongs to those households. In the area. the majority of households either got their land through allocation by the Village Headmen or through inheritance.

Despite the rise in household population density (particularly in some villages), the commons remained pretty much untouched, with settlements (even in the events of '*madiro*') confined to land within the settlements.[4] Due to land exhaustion and relocation of households to other areas (the resettlement farms and the new cotton frontier in the mid-Zambezi Valley), there was always available land for fields and settlements.

The valley community also controlled parts of the watershed area, which they shared with the Shangani River Valley communities, and some households had fields on this land, which was largely a grazing zone. The

[4] Acts of "*madiro*" or freedom farming became popular in Zimbabwe during the war of independence.

availability of the Sotane Ranch, particularly after 2010, further expanded land accessible and available to households into the former commercial ranch.

Social Dimensions of Farming

In the case study, tradition and prior experience continue to inform households' farming practices. Tradition in the form of rituals inherited from "original" settlers of the land and those practiced at places of origin (Alexander et al., 2000), also shaped the way households and society interacted with the natural environment, and their perspectives and responses on or to natural forces such as droughts and pests. Not surprisingly, traditional rituals such as environmental cleansing (*ukwebula ingxoza*) (Thebe, 2017b) played a significant role in intervention to droughts or dry spells. The participating households, as part of the wider society, believed in these society-wide rituals, and observed them without fail.

Appeal to the Supernatural

In this society, rains were conceptualised as gifts from some supernatural being that commanded respect from all those living on the land, and lack of rains were often interpreted as the anger of the spirits. Accordingly, there were certain "dos" and "don'ts" that guided farming practices to avoid angering the spirits of the land, and since some violations and natural events would still occur, rituals were performed to cleanse the environment.

The environmental cleansing ritual was associated with rainmaking, and was often performed just before the rain season commenced or when dry spells persisted for long periods. The farmers that participated in the study associated these rituals with local rain spirit mediums (*amawosana*), who were the spiritual beings and constituted the rainmaking mediums of the area. As noted in Thebe (2017b:2):

> The ritual is initiated by community elders and led by *amawosana* (spirit mediums) at the local *motolo* (rain shrine), and performed mostly by men through a ritual known as *ukwebul' ingxoza* ('debarking a tree').… The actual cleansing is focused on specific elements, such as exposed animal bones, cobwebs, nests of certain species of birds, and trees that were struck down by lightning, which are gathered and destroyed by fire.

In the past, these were perennial practices which, all other solutions failing, the request would be taken to the Njelele shrine in the Matopo Hills in Matabeleland South Province. In the area, this imperative had been used during every dry year in the past. However, respondents indicated that the practice was no longer used since the chief '*iwosana*' spirit medium had died. However, the cleansing of the forest remained as an option, although it had been rarely used. According to informants, this ritual was only performed once in the past ten years.

The absence of an '*iwosana*' spirit medium and failure to cleanse the forest were however cited as the reasons behind persistent rain failure as highlighted in the following:

> There is a lot in this forest, which scares away rains. These should be removed, but in this society there are no men, and since .....died, there is no '*wosana*' to lead the people. They also say that the spirit at the Njelele shrine stopped talking following invasion by the army.... rain will not fall in Matabeleland. The spirits are angry; they must be appeased. We have sent people to the Njelele shrine with seeds, but the spirit has forsaken us. The seeds are not blessed, there is no harvest. You can see for yourself.[5]

Despite these fears, the taboos associated with these supernatural beings were still being upheld, and sanctions applied where violations occurred, to appease the spirits. For example, people were forbidden to work their fields on Wednesdays, after thunderstorms and after rains fall following a long dry spell. Some of the households had been sanctioned in the past for violating these farming practices. One of the household heads indicated that his sons had gone to the field following a storm, and was reported to the Village Headman, and the village council ordered him to pay a goat to appease the spirits.

Environmental Knowledge

Farm households also understood their environment, and used this knowledge to plan farming activities. What was interesting was the farmers' knowledge and understanding of weather patterns: when the rains should fall?; which rains signified the planting season?; the months of calm rains?; months of dry

[5] Interview with Gogo MaMpofu in Gumede, January 15, 2018.

spells?; months of good rains?; when there was likely to be a drought and good rains?, etc. In their calendar, rains normally set-in during the month of August, and these were not planting rains.

The agricultural season began in late October, heavy rains were expected in December, which was also regarded as the month for weeding. In their calendar, continuous rains were expected from October to December, and while the growing season spread from October to May, planting was confined to the first three months. Any crops grown after December were considered late crops, and according to one of the community elders, "they sometimes failed to fully mature".[6]

Discussions on the rain season, rains and droughts indicated a heavy reliance on natural elements and incidents as forecast on the rain season ahead, which in turn guided behaviour and actions. These included the appearance or position of certain stars, the blooming of certain flowers, and the presence of certain birds or insect species. For example, the sound of a particular rain bird signified the coming of the rains and farmers would start preparing for the rainy season.

Farmers also said they could predict the amount of rains through the heat of the sun, and in their world a very hot sun signified good rains. Over the years, this knowledge had assisted farmers' decisions on when to plant, when to stop, what crops to grow and where? While rain patterns would change from year to year, this knowledge allowed households to produce crops, which would grow under particular conditions. However, such knowledge was proving to be a constraint on farming decisions by preventing farmers from taking risks, which in turn constrained initiative.

Farming Practices

While each household developed its own agricultural method based on the availability of resources and constraints, there were many similarities among households, which conformed to the elements described below:

Use of animal draft: Tilling involved animal draft, pulling a plough controlled by an individual. Only a few households could mobilise an ox-span since the cattle stock had been decimated by droughts. Only twelve households used cattle as draft power, and none had a span of oxen. A larger part of households was using donkeys, and since households had large fields, large

[6] Interview with Nkambule in Gumede, June 29, 2017.

parts remained uncultivated as a donkey span is relatively slower than an ox-span. In the study, 18 of the households had failed to cultivate all their fields during the past 10 years.

During the period following the occupation of the grazing zone (see Thebe, 2017a) and subsequent development of fields in the Sotane Ranch, donkey spans were even used on the heavy black clay soils, although donkey draft power is not suited for such soils. This challenge was particularly serious during heavy rains, forcing households to reduce the depth of the plough. This had negative results as revealed below:

> If the depth of the plough is reduced, it prevents the roots of the crops from penetrating deeper, which means that in the event of inter-seasonal dry spells the roots of the maize crop are exposed to the heat of the sun and the crops ultimately die through baking.[7]

On average, each household used a span. In some cases, a span was used by two or three households pooling resources together. Where agricultural resources were shared, fields were worked in alternate days.

Given the prominence of dry spells, only small areas could be cultivated, leaving large parts of the fields fallow. There was no hand-hoeing of fields. The plough remained the main tool and households with no ploughs could still access them through a system of resource-pooling, where they offered their labour to those with ploughs and draft power.

Since labour was a scarce resource, these networks of resource mobilisation were common, particular in households where there were a number of claimants to a single span. There was, thus, excessive demand on the available draft power, except for the few that could mobilise more than one span. However, such households had become very scarce in the 2000s following the disintegration of extended households (see Thebe, 2014).

Failing to Adapt or Doing What Is Familiar?

Preference for Maize

Households face the constant threat of crop failure due to climatic variability, resulting in food insecurity. In the past, they produced drought resistant crops.

[7] Interview with Ndlovu in Gumede, December 28, 2017.

Households with fields on red clay soils produced sorghum, while those situated on sandy soils grew pearl millet. However, such crops, although still produced by some households, were no longer popular and some households had completely discontinued their production. Indeed, households had adopted maize, which was grown by every household irrespective of the quality of the land.

The popularity of maize was impacted by perennial harvest failures, which led to the loss of traditional millet and sorghum varieties. The option was to purchase seeds from outlets that sold inputs. The production of maize was also a by-product of the livelihood path of many households, particularly the availability of the wage. Households' access to the wage gave them the capacity to purchase inputs, and maize seeds were the most readily available than sorghum and millet varieties.

However, the bulk of the land was marginal for the production of maize, and without enhanced soil fertility, crop failure occurred frequently. The production of maize by households, while convenient, was also highly dependent to an important degree on the application of organic manure.

Significantly, the droughts prevalent in the 1990s and 2000s had wiped out the cattle herd (the major source of organic manure), leaving households holding onto donkeys whose organic matter was rarely used for soil fertility. Consequently, the loss of cattle meant that households lost a key component in soil fertility, as well as in soil water retention.

During the rain seasons, households intercropped maize with a variety of vegetables including pumpkins and melons. The cropping practices that were adopted by the households were deeply rooted in the traditional farming system, as households have always practiced intercropping. For the poor households, access to maize seeds was mediated not only by their ability to preserve seeds from the previous season, but also, the development of social networks, although charity organisations and the government also distributed inputs to the poor.

Drought resistant seeds, such as sorghum, were also distributed as humanitarian aid by these organisations, but very few households committed land to the production of sorghum, and the few that produced sorghum did so only on a smaller scale, and mainly for sale or to brew traditional beer. In 2018, there was only a small number of households that had dedicated land to the production of millet, and these achieved a decent harvest in a context where other crops failed.

> The Nyamande household, for example, had fields a number of fields located on different types of soils in the settlement, and had also developed another field in the occupied former commercial ranch. While the household grow a combination of crops such as hybrid maize, melons, pumpkins and sugar reeds on fields located on clayey soils, pearl millet was also grown on fields that were located on sandy soils. In June, the other crops had all failed, but the pearl millet crops survived. The household achieved a harvest of 10 bags (an equivalent of 800kg). This was mainly used as chicken feed, or sold to other households since pearl millet was not a stable.

Before the 2017 season, the area experienced successive below normal rains, which had resulted in the failure of the maize crops. The number of years of less than optimal rains, 4 years, was among the longest experienced in the area since the 2002 to 2005 seasons, making it a particular lesson for maize farmers. However, even with such experience, households failed to shift to alternative short duration cereal crops such as short maturing sorghum varieties during the 2017 and 2018 seasons, preferring instead, to retain maize as the major cereal crop.

As shown from the Nyamande household case, during the two agricultural seasons (2017 and 2018), following the drought years, land was allocated not only to maize but also to other crops such as sugar reeds, which traditionally is a major crop in rural small-scale farming. Households that originally grew these crops in combination with sorghum, were combining these with maize. This, then, partly explains the general preference for certain crops, as also seen from the following:

> Sorghum and millet do not generate money, and have no place in people' social and economic life. Very few people still consume sorghum and millet here. But, with maize, you can sell and generate "petty cash" and also consume. The money one gets from the sale, they can cover other demands including other consumption items. Why would one grow a crop that he can use? Some people grow it for their chicken, but that is a luxury.[8]

However, others explained that their lack of interest in sorghum was because the consumption quality of the short maturing variety was highly compromised and the taste was horrible. Yet, they could not produce their traditional sorghum variety because of the short rainy season. This, then, resulted in maize being practically the only consumption cereal produced by these households.

[8] Interview with MaNkonjeni in Gumede, January 11, 2018.

Rigid Rain Calendar

Although present experiences with rain trends had changed from those of the past, farm households responded in familiar ways. In the 2015/2016 agricultural season, long dry spells between November and January destroyed all crops, with most households failing to achieve any harvest. In 2017, rains were normal and the farming season was normal. However, in 2018 heavy rains were only recorded in December.

Most households planted in November following rains in October, although there were no rains in August to mark the beginning of the rain season. The period after November was very dry and crops that were grown died of water stress. The rains finally arrived in late December, and as per tradition, farmers considered them to be late rains, and failed to take advantage of the rains, opting instead to plant the usual late crops.

But the good rains continued well into the month of February. One of the farmers recalled, "… people had packed their ploughs. Everyone expected a drought year".[9] Another farmer concurred, "I could not replant in December because January is known for its lack of rains …and crops planted in January rarely mature".[10] However, crops planted in January matured and the few farmers who had taken the risk and planted maize achieved a good harvest. One farmer noted, "The late rains had basically pushed the planting season forward, yet we based their decisions on past experiences".[11]

Climatic uncertainty – which is a manifestation of changing climate – resulted in agricultural failure and food insecurity for many households. Yet another part of the story is a new threat to the natural environment, including new land clearances and cultivation of vulnerable land by desperate farmers. In the study area, as in much of the '*gusu*' region in the 2000s and 2010s, new land clearances were witnessed as households sought solutions to failed harvests by clearing more land for fields.

The participants indicated that they acted out of pressure to improve crop production by finding new fertile land. For example, since 2010 households have been on a crusade, developing fields on every available space. First, they reclaimed old fields on the watershed; second, they occupied land on the river valley after 2012, and established fields in the ranch in 2017 (Thebe, 2017a).

---

[9] Interview with Khulu Mpofu in Gumede, July 15, 2018.
[10] Interview with Sibanda in Gumede, July 19, 2018.
[11] Interview with Baba Madonsela in Gumede, July 26, 2018.

The search for fertile land became a principal vehicle through which households responded to climatic uncertainty and harvest failure. People's understandings of the state of agriculture were based on early experiences of farmers who had fields on particular types of soils. Farmers cited incidents of land self-provisioning in 2012 on the river valley and the harvest achieved by land occupiers.

One female farmer indicated that she was forced to establish fields in the former commercial ranch in 2017 after seeing the success of farmers who had fields on the river valley, and the failure of crops in the exhausted settlement area fields in 2015. She was emphatic, "those with fields on the river valley delivered scotch cart after scotch cart of grain, and we were forced to buy grain from them".[12]

Her sentiments were widely echoed by other farmers who had also developed fields in the abandoned ranch, even though they had fields in the settlement area, who agreed that "black soils offered better prospects for a harvest than the '*gusu'* and mixed soils in the settlement area".[13] In fact, a few of the households held fields in the settlement area, river valley and the ranch. However, these households were concentrating efforts on the new fields at the expense of fields in the settlement area, which further exacerbated their food security situation since the heavy black soils required lots of rains. In 2018, for example, crops in the former ranch were completely destroyed by a herd of stray elephants, leaving households with virtually no harvest.

In early 2017 and 2018, some households had not grown any crops. Historically, this would have been households with access to remittances from the city, who treated farming as a part-time activity that only supplemented the wage.

However, in 2018, even households that were known for their agricultural acumen, and had often produced surpluses had no crops. They indicated that this was not because of lack of commitment, but they were disadvantaged by the soils and rain trends more than others. As one female head of household explained "we occupy '*iphane*' (mopane land) of red clay that requires rain, and is difficult to cultivate when rains are erratic….what do you do, waste seeds?".[14]

While the link between the failure to cultivate land and changes in climatic conditions cannot be easily drawn, rain failure had forced households to

---

[12] Interview with Masuku in Gumede, July 13, 2018.
[13] Interview with Mgujini in Gumede, July 09, 2018.
[14] Interview with Gogo MaNkabinde in Gumede, June 30, 2017.

abandon fields on red clay in favour of the virgin black soils along the river valley and in the former ranch. According to household sources, these soils were failing to yield, and crop failure due to water stress had increased in the recent past.

In the 2011/2012 season, the water stress caused by less than optimal rains destroyed the entire maize crop that was planted in December, and households gradually began reducing areas put under cultivation. In January 2018, large areas of land remained uncultivated and some fields had grown into thick bushes.

Conclusion

This chapter presented evidence from a study of small-farm households in the semi-arid Lupane District, north-western Zimbabwe, to highlight the significance of history, past experiences and certain cultural factors in understanding how small-farm households relate and respond to changes in their environment.

While for many involved in the climate change debates, emphasis has been placed on developing adaptive capacity of human communities and societies for them to withstand shocks and stresses associated with varying climatic conditions (see, Smit and Wandel, 2006), evidence from the study points to striking continuities that exist between households' past experiences (for example, the use of the plough, agricultural extensification and intercropping) and present practices, even in a context of changing weather patterns.

Tradition, and in particular, certain rituals inherited from "original" settlers and those practiced at places of origin (see Alexander et al., 2000) – alongside prior experiences – emerged as central determinants in how farmers interpreted and responded to changes in weather patterns. Moreover, farming strategies were revealing, suggesting that households continued to favour the traditional farming methods, despite the promotion of new agriculture technologies and drought tolerant crops by the government and NGOs.

Viewed as a whole, these households' responses to environmental changes were the outcome of a multiplicity of what others have referred to as "cultural dimensions of lives and livelihoods" (Adger et al., 2013:112), operating in combination, and which indicates the complexity of the adaptation question.

The influential effects of these factors at the level of household was amply illustrated by the fact that households often based farming decisions on their local knowledge, and more often, the unquestioned wisdom that had proven to be a valuable resource in their farming practices for generations. In the particular case of this society, there certainly was no emancipatory release from these cultural dimensions of farming and their influence over how people interpreted and saw natural events, as many of the rituals, including the belief in cleansing ceremonies, remained unquestioned.

For this reason, adaptation and households' responses occurred and were mediated by these socio-cultural limits, rather than anything outside. Drawing from Jim Scott (1986:30), "the goal, after all, [of the farmers' responses] is ….to survive - today, this week, this season - within it".

Significant implications emerged for any policy for climate change adaptation. First, the introduction of technologies that are at odds with societal values are unlikely to succeed. Efforts by government to introduce compound fertilisers have been less successful compared to recommendations to use organic manure, and the plough remained a popular agricultural tool for households despite the promotion of minimum tillage technologies by both government and non-governmental organisations.

Second, the adoption of drought tolerant varieties may be the most rational and efficient strategy to overcome farmers' vulnerability to food insecurity, but such crops have zero value to households since they do not form part of the diet, and in the case of this community, a shift towards the production of these varieties would upset the consumption culture that had developed with the adoption of maize. Farm households would rather produce maize even at the risk of food insecurity because maize is a central part of their identity than these new variety drought tolerant cereals.

References

Adams, Richard M., Brian H. Hurd, Stephanie Lenhart, and Neil Leary. 1998. Effects of global climate change on agriculture: an interpretative review. *Climate Research* 11:19–30.

Adger, Neil W., Jon Barnett, Katrina Brown, Nadine Marshall, and Karen O'Brien. 2013. Cultural dimensions of climate change impacts and adaptation. *Nature Climate Change* 3:112–117.

Adger, Neil W., Suraje Dessai, Marisa Goulden, Mike Hulme, Irene Lorenzoni, Donald R. Nelson, Lars Otto Naess, Johanna Wolf, and Anita Wreford. 2009. Are there social limits to adaptation to climate change? *Climatic Change* 93:335–354.

Alexander, Jocelyn, JoAnne McGregor, and Terrence Ranger. 2000. *Violence and memory: one hundred years in the 'dark forests' of Matabeleland*. Oxford: James Currey.

Andersson, Jens A. 2007. How much did property rights matter? Understanding food insecurity in Zimbabwe: a critique of Richarson. *African Affairs*106 (425):681–90.

Batterbury, Simon P. J. 1996. Planners or performers? Reflections on indigenous dryland farming in northern Burkina Faso. *Agriculture and Human Values* 13(3):12–22.

Batterbury Simon P. J. 2008. Anthropology and global warming: the need for environmental engagement. *The Australian Journal of Anthropology* 19:62–68.

Bird, Kate, and Andrew Shepherd. 2003. Livelihoods and chronic poverty in semi-arid Zimbabwe. *World Development* 31(3): 591–610.

Baudron, Frédéric, Jens A. Andersson, Marc Corbeels, and Ken E. Giller. 2012. Failing to yield? Ploughs, conservation agriculture and the problem of agricultural intensification: an example from the Zambezi Valley, Zimbabwe. *The Journal of Development Studies* 48(3): 393–412.

Brondizio, Eduardo S., and Emilio F. Moran. 2008. Human dimensions of climate change: the vulnerability of small farmers in the Amazon. *Philosoph. Trans. Royal Soc.* 363:1803–9.

Brown, Kathyn S. 1999. Climate anthropology: taking global warming to the people. *Science* 283:1440–1441.

Brown, Donald, Rabecca R. Chanakira, Kudzai Chatiza, Mutuso Dhliwayo, David Dodman, Medicine Masiiwa, Davison Muchadenyika, Prisca Mugabe and Sherpard Zvigadza. 2012. *Climate change impacts, vulnerability and adaptation in Zimbabwe.* IIED Climate Change Working Paper 3, International Institute for Environment and Development, London.

Crate, Susan A. 2011. Climate and culture: anthropology in the era of contemporary climate change. *Annual Review of Anthropology* 40:175–194.

Falk, Moore S. 2001. The international production of authoritative knowledge: the case of drought stricken West Africa. *Ethnography* 2(2):161-1 89.

Food and Agriculture Organisation (FAO). 2005. *The state of food and agriculture*. Rome: Food and Agriculture Organisation.

Kandji, Serigne T., Louis Verchot and Jens Mackensen. 2006. *Climate change climate and variability in Southern Africa: impacts and adaptation in the agricultural sector.* World Agroforestry Centre (ICRAF), United Nations Environment Programme (UNEP), Nairobi.

Kelly, Mick P., and Neil W. Adger. 2000. Theory and practice in assessing vulnerability to climate change and facilitating adaptation. *Climate Change* 47:325–352.

Mazzeo, John. 2011. Cattle, livelihoods, and coping with food insecurity in the context of drought and HIV/AIDS in rural Zimbabwe. *Human Organization* 70(4):405–415.

Mubaya, Chipo P., Jemimah Njuki, Eness P. Mutsvangwa, Francis T. Mugabe, and Durton Nanja. 2012. Climate variability and change or multiple stressors? Farmer perceptions regarding threats to livelihoods in Zimbabwe and Zambia. *Journal of Environmental Management* 102:9–17.

Mutekwa, Vurayai and Samuel Kusangaya. 2006. Contribution of rainwater harvesting technologies to rural livelihoods in Zimbabwe: the case of Ngundu ward in Chivi District. *Water SA* 32(3):437–444.

O'Brien, Karen L., and Johanna Wolf. 2010. A values-based approach to vulnerability and adaptation to climate change. *WIREs Climate Change* 1:232–242.

Orlove, Ben. 2005. Human adaptation to climate change. *Environmental Science Policy* 8:589–600.

Pielke, Roger A. J. 1998. Rethinking the role of adaptation in climate policy. *Global Environmental Change* 8:159–170.

Rayner, Steve. 1989. Fiddling while the globe warms? *Anthropologv Today S (6):*1-2.

Rayner, Steve. 2003. Domesticating nature: commentary on the anthropological study of weather and climate discourse. In *Weather, Climate, Culture*, Strauss, Orlove B (eds). Oxford, New York: Berg Publishers.

Scott, Jim. 1986. Everyday forms of peasant resistance. *The Journal of Peasant Studies* 13(2):5-35.

Thebe, Vusislizwe. 2014. New social formations, livelihood transition and food insecurity in workerpeasant communities. *Journal of Developing Societies* 30(1): 45–67.

Thebe, Vusilizwe. 2017a. Legacies of '*madiro*'? Worker-peasantry, livelihood crisis and '*siziphile*' land occupations in semi-arid north-western Zimbabwe. *Journal of Modern African Studies* 55(2):2001–224.

Thebe, Vusilizwe. 2017b. Defenders of the woods? Women and the complex dynamics of a workerpeasantry in western Zimbabwe. *South African Review of Sociology* 48(2):1–17.

Thebe, Vusilizwe, 2018. 'Men on transit' and the rural 'farmer housewives': women in decision-making roles in migrant-labour societies in north-western Zimbabwe. *Journal of Asian and African Studies* 53(7): 1118–1133.

Thomas, David S. G., Chasca Twyman, Henny Osbahr and Bruce Hewitson. 2007. Adaptation to climate change and variability: farmer responses to intra-seasonal precipitation trends in South Africa. *Climatic Change* 83:301–322.

Biographical Sketch

Vusilizwe Thebe is a Professor of Development Studies and Head of the Department of the Department of Anthropology, Archaeology and Development Studies at the University of Pretoria. His research interests include migrant labour societies and their transformation: the worker-peasant social dynamics, livelihood transition, migrant labour, remittances and their movement, gender dynamics, the youth and women employment question, and agrarian transformation. He has published numerous articles in international journals on social dynamics in former migrant labour societies, migration and the youth question.

Chapter 2

The Significance of Wage Employment and the Enduring Unemployment Crisis and its Effects on Peasant Farming in Zimbabwe

Moffat Chiba, PhD
Department of Archaeology, Anthropology and Development Studies, University of Pretoria, Republic of South Africa

Abstract

This chapter explores the significance of wage employment to rural livelihoods and how the enduring formal unemployment crisis, witnessed since the 1990s, has affected peasant agriculture. Using 30 migrant labour households' historical accounts in two rural settlements in the Shamva District of Zimbabwe, evidence shows that wage employment, especially in the context of climate change, has enhanced rural households' agriculture, and maintained a dualised connection between the rural and urban frontiers. However, the enduring formal unemployment crisis has destroyed rural households' social, financial, physical, natural and human capitals and hence smallholder agriculture and is currently causing a social protection crisis, especially in old age. There is, thus, need, on the part of the government, to create formal employment through influencing fiscal and monetary policies to broaden the industrial base. The proper management of skills training by managing schools and tertiary institutions responsibly may enable skills to tally with available formal employment opportunities to restore the pre-1990s worker-peasant existence.

Keywords: peasant agriculture, rural households, unemployment crisis, wage employment, Zimbabwe

In: Agricultural Policy
Editor: Youssef Lee
ISBN: 979-8-89113-720-2

Introduction

Wage employment, an economic activity that comprises all activities where a person is hired and performs labour in return for a fixed monthly salary, an hourly or daily wage, or a fixed piece wage rate (Van Hoyweghen, Van Den Broeck and Maertens 2018), is generally a significant livelihood portfolio in both high- and low-income countries in an international economy that has wholly become monetised. Writing over a decade ago, a World Bank report recognized that wage employment is an important tool for poverty alleviation (World Bank 2008) and improves household welfare (Van Hoyweghen, Van Den Broeck and Maertens 2018). In some contexts, wage employment, is harnessed as a short-term fix for seasonal harvest variation, for example ganyu employment in Malawi (Van Hoyweghen, Van Den Broeck and Maertens 2018). In other settings, it is widely recognised as a source of additional income that increases households' standards of living (Fink et al., 2014).

Notwithstanding this widely recognised role of the wage in the survival matrix of humanity, most economies of the world are in the current entangled in complicated archaeologies of economic crises including high levels of formal unemployment.1 More often, periods of economic affluence are often followed by periods of economic downturn (Fleşer & Dobre-Baron 2010). Achy (2011), writing of Tunisia, recognised that Tunisia is facing high rates of youth formal unemployment, a large number of marginal jobs, increasing income inequality, and substantial regional inequalities. Globally, increasing levels of food insecurity owing to the disruption of the Global Food Supply Chains by the enduring COVID-19 epidemic are well-documented (Van der Ploeg 2020). Besides, a greater majority of people in late capitalist economies are experiencing problems of social protection in old age due to lack of gainful employment during their working lives (Puentes, Contreras and Sanhueza 2007).

While a lot has been learnt from this literature (for example Mhazo and Thebe 2021), here, I pay particular attention to the significance of wage employment to the livelihoods of smallholder farmers considering that

[1] As of 2014, average general unemployment rate in the SADC region stood at 40% (Malepe 2014:3). The rate of unemployment in Zimbabwe differs. The archbishop of New York places it at 90% (www.bbc.com/news/business-42116932) whilst others places it at 5% (https://tradingeconomics.com). Unemployment generally refers to the number of people between the ages of 15 and 60 willing to work but are failing to get a job (www.bbc.com/news/business-42116932). The number of employed people declined from 20.9% in 2004 to 15.9% in 2014 and the majority of those who should be working and accounting for 74.7% in 2014 were in vulnerable employment (Kanyenze 2018:115).

dualised livelihood portfolios have emerged since the 1950s when migrant workers have been straddling between the rural and urban frontiers (Bernstein 2004), but this dual existence of the worker-peasant seem to have abruptly come to an end due to the ongoing formal unemployment crisis that has begun to take a huge toll since the 1990s (Thebe 2014,45). I, thus, examine the formal unemployment crisis that manifested itself from the 1990s to the current using the Zimbabwean case. The chapter contends that although wage employment is a central livelihood portfolio among smallholders, there is an ongoing formal unemployment crisis that started from the 1990s; a period that marked the beginning of cases of unemployment (Stoneman 1991). Thus, "the turn-around policies from the 1990s together with the deepening crisis since the turn of the second millennium, left households without the secure urban support, as the urban sector could barely support the rural sector" (Thebe 2014). The decline of the Zimbabwean worker–peasantry was predicted by Cousins, Weiner and Amin (1992) and in its place, they foresaw the emergence of a "lumpen semi-peasantry," and a "rural petty bourgeoisie." While this may have been true at the time, the rural petty bourgeoisie has not really replaced the worker–peasantry, instead, dominant in rural areas is a new class of "marginal farmers," with no or limited support from the urban sector (Thebe 2014,46-47). The argument I advance here is that this crisis has a strong bearing on rural households' livelihoods including their agricultural activities. Accordingly, I contend that this is the fitting starting point for exploring the crisis of unemployment.

This study, thus, becomes undeniably significant considering that globally, country-specific economies are experiencing economic dynamics in the midst of the COVID-19 pandemic (Van der Ploeg 2020), a novel viral disease, first identified in Wuhan, China (World Health Organisation 2020) and later spread to the other parts of the world. Currently, it has affected negatively Global Food Supply Chains (Van der Ploeg 2020), meaning that governments need to revamp their territory-specific industries including their food production systems to enhance standards of living of their people. Besides, 'the speed and measure of the changes coming about by the fourth industrial revolution are not to be ignored. These changes will bring about shifts in power, shifts in wealth, and knowledge. Only in being knowledgeable about these changes and the speed in which this is occurring can we ensure that advances in knowledge and technology reach all and benefit all' (Xu, David and Kim 2018, 90). Currently in Zimbabwe, educational institutions have sprouted at a period of rapid deindustrialisation. Graduates are being channelled out of tertiary institutions with a gloomy picture of employment prospects (Mhazo and Thebe 2021). This study, thus, absolutely becomes relevant as the issue of unemployment, especially among the youth, has emerged as a major policy concern (Mhazo and Thebe 2021), not only in emerging economies but in economically advanced countries as well.

In this chapter, I start by advancing my conceptual framework, which I use to anchor the study before discussing my materials and methods. After this, I examine, first, the emergence of wage employment in Zimbabwe and thereafter assess the significance of wage employment before discussing different schools of thought that have shaped debates on the post-2000 economic crises to place the study in current scholarship before discussing the formal unemployment crisis faced since the 1990s. I then explore the implications of the formal unemployment crisis on rural households' agriculture and livelihoods before discussing available survival options and their implications. Finally, I discuss the findings and conclude.

Conceptual Framework

I used ideas of the Sustainable Livelihoods Framework (SLF) to problematize my argument since my hypothesis was that macro-economic policies (Structural Adjustment Programmes of the 1990s, the awarding of $50 000 handshake to the war veterans, involvement of the country into the Democratic Republic of Congo War during the late 1990s, continued corporatist control of the exchange rate and the post-2000 land redistribution exercise), have generated high levels of formal unemployment among the worker-peasants, affecting their agricultural production processes including their asset endowments.

The Sustainable Livelihoods Framework is often identified with various scholars including Scoones (1998) and places ordinary people at the centre of livelihoods studies. A livelihood entails capabilities and strategies needed to sustain a living. It is sustainable if it can exploit its capabilities using available assets to withstand shocks and stresses (Ashley and Carney 1999). The approach places emphasis on the following 5 important elements in people's livelihoods: The vulnerability context, assets, transforming structures and processes, livelihood strategies and outcomes (Scoones 1998). The Vulnerability context is where people live and survive using available assets which include financial, social, natural, human and physical capitals (Ashley and Carney 1999). These determine the ability of people to enjoy well-being (Ellis and Biggs 2001). Financial capital relates to the ability of people to access income either through wage employment or accessing loans from financial institutions (Ashley and Carney 1999). Social capital relates to the strength of relationships between and among community members. While natural capital relates to assets like land, physical capital relates to tangible assets like livestock, farming implements and equipment. Human capital refers to skills and knowledge that a particular household or community possesses through its members. Transforming structures and processes include policies, organisations and institutions that either support people's access to assets or expose them to risk (Ashley and Carney 1999). These include government ministries and financial institutions which give loans to people including laws that legalise rights to property ownership.

The SLF, thus, allowed me to ascertain how the macro-environment, through the generation of formal unemployment (Transforming structures and processes), have affected the functioning of smallholder farmers' five capitals (social, human, natural, financial and physical) and hence their agricultural production processes and general livelihoods.

Materials and Methods

While it was my argument that wage employment is a significant livelihood portfolio globally, my focus was confined to the Shamva District in Mashonaland Central Province in Zimbabwe, lying approximately 90 kilometres to the north-east of Harare, the country's capital. I focused on two villages-Bushu Communal Area and Mupfurudzi Resettlement Scheme. The villages were carefully chosen not for comparing, but to generate enough cases to qualify the argument that the wage was important among rural households, but a formal unemployment crisis has disturbed general livelihoods. Besides, I had had prior experience with working in the villages when I was carrying out field-work for a master's degree. Furthermore, the villages had a well-defined history of migrant labour, where combinations of hoe-wage or wage-hoe have typically been practiced (See also Bernstein 2004). However, in the current, households' life historical accounts constitute a tale of sad news concerning the serious formal unemployment crisis that these two villages have experienced. Many households' members have resultantly diversified their livelihood activities into illegal gold mining and haircutting, activities that typify disguised unemployment. Considering these factors, the two villages provided fertile ground for my analysis.

Similarly, evidence show that rural households were employed in diverse sectors of the industry before the 1990s. Of all the 30 participating households, virtually each had a member that had been engaged in migrant labour. These members have been engaged in wage employment in different sectors of the economy (see table 1 below).

Table 1. Number of households in Wage employment before and after 1990

	Before 1990	After 1990	Total
Bushu	25	10	35
Mupfurudzi	24	9	33
Total	49	19	68

Source: Field-work 2017-2018.

Bushu is an old communal area established in 1916. It is home to indigenous Shona ethnic groups, mostly of the Mazvimbakupa lineage, under the village headship of Bushu. According to village sources, the lineage arrived in the area around 1916. Bushu is fairly populated, with an average household size of 4, and is located in a basin-like valley about 9km to the north of Shamva Growth Point, with the Zambayamba hill range to the north, which runs from the east to the west and to the south, the Kajakata Hills forms the border with Kambiri Village. To the west, it borders the Nyamahumbe Resettlement Scheme, whilst to the east, it is bordered by the Shamva-Madziwa Highway, which provides a gateway to the Madziwa Mine and Teachers' College.

The other study village, Mupfurudzi (Dombojena), is a Fast-Track Land Reform village established in 2004. It lies about 23km to the north of Shamva Growth Point, bordered by Gono and Kanyemba Villages to the south and by Mupfurudzi River to the north, separating it from other resettlement schemes like Mutoramhepo, Mupedzanhamo and Murindagomo. To the east, it is bordered by the Princip Irrigation Scheme, while it acts as a gateway to Madziwa Mine and Teachers' College from the Shamva Growth Point. The distance between Mupfurudzi and Bushu is approximately 14km. In this resettlement scheme, households' land sizes are relatively small and averaging about 3.5 acres per household. I included this village owing to the diverse cultures that converged in fast-track farms at the implementation of the FTLRRP (Matondi 2012) to get a holistic picture in my analysis.

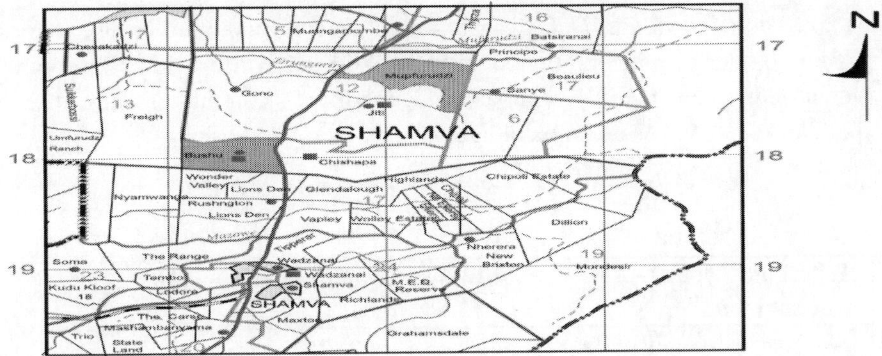

Map of Bushu and Mupfurudzi-Dombojena.

In these two different communities, I conducted an ethnographic study during the period between November 2017 and January 2019, with a focus on 30 households (15 from each study site). These households were visited on a number of times, utilising a system of protracted discussions, conditioned by my question schedule but simultaneously tolerating research participants to get deeper into an assortment of subjects on their lived experiences.

Before fieldwork, I obtained research ethics approval from the University of Pretoria and the Office of the President and Cabinet, Mashonaland Central Province, Zimbabwe, in July 2017. I relied on rural households' life histories in data gathering, which enabled me to understand households' experiences including their interpretations, viewing them as prisms through which to solicit meanings that they attach to their lived practices (Symon and Cassell 2004). I, thus, implemented a case study method where I visited households on a number of times to build comprehensive households' life history accounts on different aspects related to the economic aspects of their past and present livelihoods with the help of an interview guide. This technique was backed by participant and non-participant observations, which were particularly useful in spotting the present employment or unemployment situation of households. I also backed these tools with informal discussions involving over 100 men, women and children with diversified rural backgrounds including past literature to provide adequate evidence for the study.

I chose households that had a history of combining wage employment and rural farming. This criterion was appropriate in understanding the significance of wage employment within the smallholder farming sector (See table 2). Participants were asked for informed consent before contributing. I explained to them their role and told them that the results of the research would be disseminated in form of an article and where I intended to use names, pseudo- styles would be applied instead. Additional life history interviews were conducted in the different parts of Zimbabwe's communal lands in areas like Domboshava, Masvingo and Beitbridge.

Table 2. Number of households interviewed

Household Type	Bushu	Mupfurudzi	Total
Migrant labour	15	15	30

Source: Field-work 2017-2018.

In data analysis, I broke it into convenient themes, seeking configurations and connections (Khan 2014) and generating meaning out of the photos and text that had emerged from field-work (Denzin and Lincoln 2011). The process, thus, comprised repeated readings of the households' life histories including discussions of findings with colleagues whilst at the same time comparing the findings with earlier studies. I was, thus, able to create themes, which I then formed the thematic areas in this article.

The Development of the Wage Economy in Zimbabwe

Wage employment has a much elongated past in connection to Southern Africa's transition to capitalism (Bernstein 2004) including means of accumulation processes among rural peasants (Worby 2001). Even before the establishment of colonial rule, there has been labour migration of Zimbabweans to South African mines (Mtetwa 1976). However, the development of wage employment is often associated with the development of the worker-peasantry (Thebe 2014, 47), during the establishment of colonialism (Arrighi 1970; Bernstein 2004). This process was guided and directed by the British South Africa Company with the aid of a Royal Charter granted to Cecil John Rhodes by the British Crown in the late 1880s and these developments saw the emergence of the migrant labour economy leading to a combination of farm and off-farm livelihoods; creating hoe-wage or wage-hoe combinations (Bernstein 2004, 211), producing what has aptly been referred to as the worker-peasantry in a dualised economy. In this vein, the survival of the capitalist mode of production came to depend entirely on the labour of the African population (Duggan 1980).

To create a class of salaried employees, the British Colonial Administration in Southern Rhodesia, now Zimbabwe, passed a series of legal instruments. The 1894 Matabeleland Order-In-Council, for example, became the first segregation tool as it led to the creation of two scorched Native Reserves-the Gwaai and Shangani Reserves (Alexander, McGregor and Ranger 2000). The idea of separating Africans from land was done through land alienation and the subsequent imposition of various forms of taxes and other instruments like the Land Apportionment Act of 1930 to force Africans into wage employment. This echoes with the assertion that 'the creation of the geographical world for the natives was certainly significant in the development of capitalism in Zimbabwe, but it was equally significant in the creation of semi-proletarian communities in the reserves' (Thebe 2012). Owing to the long-established link between Africans and their land, the African peasantry had to combine agricultural activities with the wage from migrant labour. The reduction of African agriculture to subsistence levels was meant to create a class of semi-proletariat communities to serve the purpose of British capitalism and to maintain a continuous flow of labour into the mines and farms owned by the white settlers, and resultantly, a series of legal instruments including the Maize Control Act of 1934 were implemented (Bush and Cliffe 1984). Notwithstanding the importance of these plans in incorporating Africans into the wage economy, labour was often circulatory as it straddled between the rural and urban rifts and therefore there was free movement of labour (Potts and Mutambirwa 1990). Besides, the deployment of extra-economic coercion to source African labour became central (Mosley 1983).

The passing of legal instruments like the Compulsory Labour Act of 1942, made some men to escape obligatory conscriptment into the wage economy through escaping to neighbouring countries like South Africa. The enforcement of men into wage employment was, thus, made possible through the neglect of reserve African agriculture (Page and Page 1992), which was done through purposeful overpopulation of the reserves by both people and animals leading to severe ecological degradation (Prescott 1961). Though the idea of compelling Africans into the wage economy was to remove the African from reserve agriculture through the passing of various pieces of legislation including the Native Registration Act of 1936 and the subsequent Pass Laws, the connection between the African and rural land prevailed (Potts and Mutambirwa 1990). This state of affairs between wage employment and rural agriculture, thus, survived the liberation struggle well into the 1980s when the country regained its political independence from its colonial master, Britain.

This section has, thus, looked into the history of migrant labour. In the next section, we focus on the importance of wage employment since its inception to the contemporary.

The Importance of Wage Employment

Migrant labour has traditionally played a critical role in the livelihoods and survival of rural households in both colonial and postcolonial Africa. This has been particularly the case in settler and former-settler agrarian societies, where massive land appropriations had forced men to sell their labour for a wage that was not attractive. Labour migration has, thus, often been seen in a negative light, mainly due to its exploitative character, and also, for forcibly dividing families (Murray 1981). It has also been blamed for its negative effect on agriculture, by taking able-bodied men out of the rural areas, leaving agriculture as an activity entirely dominated by women, children and the aged. Despite these negative views, its contribution to rural agriculture, food security and poverty reduction has been recognised (Worby 2001). The analysis in this article was, thus, guided by these observations. It was mindful of this link between migrant labour income, agriculture and food security.

Rural households' long history of contact with the world of capitalism revealed that remittances from migrant labour have been the keystone of agricultural success and their livelihoods. Proceeds from wage employment could be used to hire labour, buy assets and agricultural inputs (Cousins, Weiner and Amin 1992). Agricultural inputs became critical elements of any agricultural process of rural households considering the progressive reduction of soil quality after centuries of continuous cultivation by the preceding generations. Resultantly, access to agricultural inputs determined households' agricultural performance and output. This importance of migrant labour in smallholder farming is reminiscent of "reserve agriculture subsidized a worker's youth, old age and family" scenario identified in Duggan (1980, 229). In other contexts, for example "during Zimbabwe's transition to capitalism, through years of coercive legislation in African occupied areas – ranging from compulsory conscription of labour (Chibaro) through the Compulsory Native Labour Act of 1942, to social engineering through the Native Land Husbandry Act of 1951 – the "hoe and wage" … "wage and hoe" … emerged as central not only in the reproduction of labour but also in the livelihoods of rural households (Thebe 2014, 45). Similarly, in this study, it was learnt that:

All the farming implements and the cattle that we were using to carry out our crop production activities have been bought from income from formal wage employment in town, particularly from my husband and children who happened, at one time, to work in the city of Harare. These included wheel-barrows, a city home, livestock and the houses that you see around this home (Gogo Mbenene 27 June 2019).

Wage employment also enabled rural households to meet their food needs. Through wage employment, households could buy food directly in the event of crop failures or used as a permanent source of food supply considering that Zimbabwe experience drought periods after every 10 years (Bird and Shepherd 2003). This was the case, since households in our sample harnessed a combination of hoe-wage or wage-hoe livelihoods (Bernstein 2004). As one respondent put it, 'My involvement in wage employment up to the early 1990s enabled us to weather our household food shortages well into the early 1990s. If someone is involved in wage employment, one can be able to withstand livelihood shocks' (Chimutengure, 10 October 2019).

There was, therefore, a strong relationship between household status, labour migration and remittances. This enabled households to command labour needed for agricultural tasks in and around the household. This was made possible because of the prevailing political and economic stability before the 1990s and as Mano, Isaacson and Philippe (2003, 1) observed, 'In the 1980s, Zimbabwe enjoyed a relatively stable macro-economic environment buoyed by international capital flows and domestic industrial and agricultural growth in excess of 3%. Inflation remained very low and the local currency remained stable …'

However, from the early to the late 1990s, varied socio-economic and political challenges gripped the country. Declining living standards and open demonstrations in cities due to inflation and unemployment became commonplace (Chiumbu and Musemwa 2012). These led to the formation of the National Constitutional Assembly, a broad-based social movement in 1998, culminating into the creation of the country's strongest political opposition since 1980, the Movement for Democratic Change in 1999 (Chiumbu and Musemwa 2012). This marked a turning point in the political and economic history of the country with the launching of the FTLRRP in early 2000s. These events had had a strong bearing on the country's economy such that at the millennium turn, a combination of economic crises including high levels of unemployment, inflation and exchange rate dynamics became common. These have generated varied debates as shown in the next section.

Perspectives on the Post-2000 Zimbabwean Economic Crisis

To understand the enduring formal unemployment crisis in Zimbabwe, it is necessary to advance some major arguments that have been forwarded to explain the post-2000 Zimbabwean crisis. Three major views have dominated the debates: the drought, internalist and the national left perspectives.

The drought argument considers natural factors like drought and extreme weather events as the main causes of the post-2000 economic crisis and food insecurity (Tadross, Hewitson and Usman 2005). To support that the post-2000 Zimbabwean crisis owes much to drought, the Food and Agricultural Organisation and the World Food Programme conducted workshops in Harare with different stakeholders and established that the cause of the 2002 food crisis was drought (Richardson 2007).

The internalist perspective or the International Left argument, views the government's Fast-Track Land Reform and Redistribution Programme (FTLRRP) launched at the millennium turn as having destroyed the economy (Richardson 2007). 145 755 peasants benefitted from the FTLRRP under A1 model, whilst 22 896 black farmers gained medium-to-large scale commercial farms under A2 model (Moyo 2007). It has been estimated that between 2000 and 2003, Zimbabwe's Gross Domestic Product collapsed by an average of 11% because of the FTLRRP of 2000 (Richardson 2005, 3). The view draws on a historiographical reading of the post-2000 Zimbabwean crises, arguing that ZANU-PF's authoritarian politics have produced and reproduced the crises through maintaining philosophies of cronyism and lawlessness (Ndlovu-Gatsheni in Chiumbu & Musemwa 2012) (for contrasting views see Moyo 2011,939).

Another perspective is the national left argument. Drawing from historical materialism, this views the Zimbabwean crises as direct outposts of imperialist deception, neo-liberalism and marginal capitalism (Moyo and Paris 2007). It considers the FTLRRP as a national project that fell victim to the global neo-colonialism agenda. The state's land and economic reforms at the millennium turn were, thus, symbols of African resistance to neo-liberalism though the state failed to garner support from its urban electorates (Moyo and Paris 2007). Whilst these perspectives are true in their own right, here we shift the angle of analysis through unravelling the macro-economic policies from the 1990s that affected smallholder agriculture.

The Unemployment Crisis after the 1990s

With the dawning of the 1990s, the formal employment situation in Zimbabwe changed. When the Zimbabwean government signed into IMF/WB Economic Structural Adjustment (ESAP) development stimulus package, the unemployment rate doubled from about 10% in the mid-1980s to about 20% in the 1990s (Marquette 1997). In 1991, net new jobs accounted for only 10% of one million school leavers since independence (Stoneman 1991, 6). About 20 000 and 50 000 retrenchments were witnessed in the public and private sectors respectively between 1991 and 1996 (Crush, Chikanda and Tawodzera 2015). Similarly, job opportunities from the non-agricultural sector significantly dropped from 939 900 to about 915 800 in 1996 (Potts 2000). Whilst ESAP had set the pace of unemployment in Zimbabwe, other 'counterproductive political decisions . . . taken from late 1997' (Brett 2005, 94) were equally destructive to the economy. These included the awarding of $50 000 handshake to the war veterans, involvement of the country into the Democratic Republic of Congo War during the late 1990s, continued corporatist control of the exchange rate and other prices. Further, the FTLRRP worsened the unemployment situation in the country (Brett 2005). Resultantly, more than 250 000 farm jobs were lost including the displacement of farm workers between 2000 and 2002 alone (Waterloos and Rutherford 2003). In the process of job losses, the informal economy grew from obscurity to unprecedented levels, further worsening government's ability to create formal employment since the taxation base had been narrowed (Mhazo and Thebe 2020). Though the economy seemed to have been recovering during the multi-currency system (2009-2013), economic woes have since returned with a severe formal unemployment crisis.

Unemployment Crisis and Smallholder Agriculture

This section provides an examination of the implications of the unemployment crisis on smallholder agriculture. It analyses the extent to which household unemployment dynamics have impacted on the agriculture of smallholder farmers. Being guided by my analytical framework, I trace the impact that the unemployment crisis has brought on households' assets including livestock, farm equipment, and urban housing. This analysis was guided by the assumption that assets are important components for any rural farming (Dorward et al., 2001; Hoddinott 2006).

1. Physical Capital

As has been highlighted in the conceptual framework, physical capital relates to tangible assets like livestock and farming implements especially ploughs. Evidence show that these have played a critical role in rural farming, but they need updating and replacement. These equipment age and once they have reached their sale by date, it is difficult for farming to be done successfully. Households with financial resources would replace equipment after 15 years, while others would replace parts. It was noticed during the study that some of the ploughs had worn-out and were lying idle under the granaries or kraals. A greater majority of the households alluded to the struggle of finding suitable farming equipment. The equipment they had relied on over the years had become obsolete.

The cost of parts was cited as a major constraint at a time of economic meltdown when most households were struggling to meet their livelihood needs, but without any source of income support. The income from wage employment, even for those who were still engaged in wage employment, was failing to cope with the costs of consumption goods. For example, Mbuya Joyce Pamire, a widow, explained that she had failed to buy the necessary spare parts needed to recondition the ox-drawn plough that was just lying idle in the open at the homestead.

Worker-peasants' households also owned houses in cities. When men lost their jobs, they either sold them outrightly because of lack of income since these houses were paid in installments which could not be provided upon the loss of income. The disposal of the urban home was a rational decision occasioned by the loss of a steady income from the wage to maintain the bond. One household head explained, 'When I got retrenched in 1993, I had joined a housing cooperative that was run by the Beverley Building Society. I was paying the residential house monthly. However, when I got retrenched from the G.M.B in 1993, I sold it' (Hwete, 16 July 2018).

Farmers of applied bent are associated with large herds of livestock and these are closely linked to wealth and power, and households that command large herds of cattle also command respect in society. They can use cattle as draft power, as source of milk, or as lobola payment. However, the period between 1980 and 2000 was characterised by droughts of differing severity (see Bird and Shepherd 2003). The 1991/1992 drought, particularly wiped out large cattle herds. Households in the study were not spared, and cattle were lost to drought. In some cases, households had to draw on the cattle herd to supplement grain. One of the households that were severely affected was the Mambo household in Mupfurudzi, which had a large herd of cattle and in the 1980s, the household had an excess of 20 herds of cattle. However, in the mid to late 1990s, the herd was reduced to ten herds of cattle. This was still a large herd by any standards, but then came the 2001/2002 droughts and the rain failure prevalent in subsequent years. Without income from wage employment, the household had to sell some cattle to meet the household's food requirements. In one of the interviews, one household head reminded me, 'After having been retrenched in 1992, I had a significant number of cattle that numbered approximately 12. However, I remember that this herd got reduced as I started disposing them to meet the daily needs of my family after facing retrenchment in 1992' (Mbenene, 16 October 2018).

Cattle were thus lost, and households that previously commanded multiple ox-spans found themselves struggling to put a span together. Other households shifted their attention to the ownership of donkey, although they still maintained herds of cattle.

2. Financial Capital

In the conceptual framework, financial capital related to the ability of people to access income either through wage employment or accessing loans from financial institutions (Ashley and Carney 1999). The wage has always played a crucial role in rural farming through the purchase of inputs. However, the period after the early 1990s was a different period. It was characterised by high levels of unemployment, job losses and a return of former migrant workers to the rural areas. Households' members lost their employment and consequently faced serious struggle to access inputs, like maize seeds and chemical fertilisers. This was worsened by the removal of government support on rural farmers in the provision of agricultural loans and subsidies through the Agricultural Finance Corporation. A fitting example was that of Gerald Kajese, who lost a job as a general employee at the Coca-Cola Company of Zimbabwe in 1998. With the loss of his job, Gerald went to his rural home, but was finding it difficult to buy agricultural inputs since he had lost income.

Agricultural inputs had become indispensable since the land in rural areas was rapidly degrading and had increasingly become a challenge in the production of crops. The gova2 soils in Bushu could no longer sustain crop production without applications of agricultural inputs in the context of an ailing economy. At the time of study, a single bag of top-dressing fertiliser (ammonium nitrate) was costing the farmers ZW$400 (400 bond/RTGS 400); a cost that could not be afforded even by a formally employed civil servant. The other space, Mupfurudzi, though a resettlement site, with relatively better soils had its own challenges. Households' members in this locality did not have a regular and reliable supply of income making agriculture an unviable economic activity. Thus, in terms of rural households' agriculture and food security, both sites had challenges and the failure to command the needed inputs were among the most worrying.

3. Social and Human Capital

The enduring formal unemployment crisis in Zimbabwe that began to take a huge toll since the early 1990s has also disturbed the social and human resources that for long had guided smallholder livelihoods. These two forms of capitals were lost when other household members left the country to look for greener pastures elsewhere. For example, Destiny, the son of Jairosi Kadhindi was retrenched from the G.M.B job in 2003. After leaving the G.M.B, Destiny went to South Africa where he is now working. However, the challenge that the household faces is that he is too far from home, and without any immediate family to support, the support from Destiny is irregular and he only comes home once every 2 years.

Households with members that were ready for the job market were similarly affected as it became extremely difficult to get a job. In rural areas, children of a certain age leave the rural area to seek job opportunities in the cities. They were often assisted by people already in the system, who would recommend them to their employers. However, after 1990, this proved difficult since these networks were either retrenched, or did not have secure jobs or had migrated to other countries, and no employer was hiring.

[2] Gova soils mean clay soil. These soils are difficult to work in the dry season, but at the same time can be very difficult to work once rainfall becomes too much.

4. Natural Capital

The triumph of capitalism and the consequent end of bipolar politics from the 1990s signified a major turning point in the history of global economic restructuring in which the capitalist ideology emerged as the victor in its long struggle with communism since the 1950s during the cold-war politics era. While the climate change phenomenon has largely been viewed entirely as both technical and bio-physical phenomenon, recent evidence show a strong connection between climate change and the current capitalist development model. Thus, the logic of capitalism especially in environments where higher levels of unemployment have become commonplace, the natural environment has constantly faced destabilisation since the 1990s. The engagement of rural households' members in other livelihood ventures after retrenchments, especially in brick-making and gold-panning, have not spared the environment through deforestation. In brick-making, trees are cut to provide energy for burning bricks. In gold mining, trees are cleared off the mining sites. We were thus informed, 'we really know that cutting down trees is bad for the environment and causes climate change. However, we need to survive. So, we engage ourselves in brick-making at least to remain alive and survive. We do not have other means of survival' (Munodawaenda, 3 February 2018).

Available Livelihood Options and Future Implications

Having either been retrenched out of wage employment or failed to get a job placement, rural households members had no other available livelihood options except turning to the informal sector. Non-farm activities are now common in Zimbabwe following the decline in employment opportunities and the meagre wage paid in both the industry and public sectors. In the study, nearly all households had people who were involved in one informal activity or another including selling some of their farm produce like maize cobs and pumpkins, while children and women would buy vegetables in bulk, for resell at the nearest business centres (Shamva Rural Town, Bushu Business Centre, Chakonda and Magetsi), particularly during off-peak seasons. Others had become involved in illegal gold mining, haircutting, pop-corning and brick making. The joining of the informal sector by many households' members means that the problem of social protection is currently in the making as the formal unemployment crisis continues to bite the economy.

Discussion

By examining the employment histories of rural households, the results show that wage employment contributed to the success of the smallholder agricultural enterprise, acted as a source of food security and reduced poverty (Duggan 1980). The analysis in this article was, thus, guided by these observations. It was mindful of this link between migrant labour income, agriculture and food security. It recognised that any loss of income from migrant labour is more likely to negatively affect agricultural processes.

However, a formal unemployment crisis, recognised since the 1990s, has negatively affected the ability of smallholders to carry out any meaningful agriculture and general livelihoods. This crisis has been set in motion, initially by the adoption and implementation of Structural Adjustment Programmes and later by other policy errors by the government, for example, the awarding of 50 000 handshakes to the War Veterans in 1997 and the implementation of the Fast-Track Land Reform and Redistribution Programme (FTLRRP) after the millennium turn (Mhazo and Thebe 2020).The long-established relationship between wage employment and rural agriculture has been delinked owing to an ongoing formal unemployment crisis from the 1990s. Using the Zimbabwean case, I argue that as the formal unemployment crisis continue unabated, smallholder agriculture is facing declining productivity and this has manifested since the 1990s (Cousins, Weiner and Amin 1992). Similarly, in Lesotho, the ability of rural households to purchase food was significantly curtailed as remittances from Lesotho migrant labourers declined owing to a weakening demand for manual labour in South African mines towards the late 1990s (Lundahl, McCarthy and Petersson 2003).

The historical accounts of rural households have testified to the far-reaching consequences of formal unemployment of their members since the 1990s. All households demonstrated that they either had a member who had been chucked out of employment since the 1990s or someone failing to locate any gainful employment in the current. Resultantly, smallholder agriculture production and general well-being of livelihoods are losing out. The five capitals (natural, physical, financial, social and human), which have been keystones of general rural livelihoods (Dorward et al., 2001; Hoddinott 2006) have been destroyed or faced unproductive proportions.

While most rural households' members have shifted to other non-formal activities, there is an ongoing social protection crisis because "self-employment … in developing economies … closely resembles disguised unemployment" (Mandelman and Montes-Rojas 2009) and a greater majority of these self-employed workers have no access to social security, severance payments, minimum wage or working condition standards" (Puentes, Contreras and Sanhueza 2007). Though the government of Zimbabwe is seriously committed to providing social protection3 through quasi-governmental bodies such as the National Social Security Authority (Chirisa 2013), informal workers are not covered by such bodies and considering that the majority live from hand to mouth, this leaves little capacity to save for old-age social protection needs. The area of social protection, thus, remains a critical area to be addressed in low-income countries like Zimbabwe.

The results in this study are significant in predicting the future of the worker-peasants in Africa, particularly in countries like Zimbabwe. Matters related to the prospects of the worker-peasantry should be matters of central concern in the current. Whilst Karl Marx's theory predicted the demise of the peasantry at a stage when they would lose their land through the creation of large capitalist farms followed by their subsequent joining of the working class (McMichael 2006) through what has aptly been referred to as the depeasantisation processes (Bryceson 2000), the results in this article demonstrate the disintegration of the long linked connection between wage employment and farming due to the ongoing formal unemployment crisis in the country. The future of rural farming, thus, becomes central in the processes of agrarian change and development and in the words of Moyo:

> The fate of the peasantry in terms of its socio-economic character and political significance under capitalism remains central to neo-colonial Southern African futures. Is the peasantry disappearing economically or becoming politically insignificant, given the emerging perception on agrarian change since the implementation of Structural Adjustment policies and market liberalisation world-wide has had a dissolving effect on peasant livelihoods (Moyo 2005: 147).

These insights into the formal unemployment crisis have emerged as a major policy concern and signify the fate of the worker-peasantry in this web of seemingly permanent crisis. This chapter has shifted the angle of analysis of the Zimbabwean crisis in general as one that has been driven by a lack of democratic space for Civil Society Groups in the country, particularly Trade

---

[3]. Social Protection are public measures to provide income security for Individuals (Holzmann, 2001). Social protection means government policies aimed at cushioning marginalised groups against the shocks and adverse effects induced by certain exogenous factors (Chirisa 2013:128).

Unions. What this entail is that democracies that are embedded within capitalist systems need to be fundamentally changed. The adoption and implementation of the Economic Structural Adjustment Programme from the early 1990s fundamentally changed the conditions of employment for workers. Whilst the Labour Relations Act Number 16 of 1985 required Ministerial approval for worker dismissal, this Act was amended in 1992 and left these matters at the discretion of the employers to decide (Raftopoulos 2006, 217). The Zimbabwean formal unemployment crisis is to date a reminiscent of Karl Marx's argument that the most definite way of making abnormal profits is to create a reserve army of the unemployed and the underemployed and clearly, the Zimbabwean scenario is one which depicts a labour force that has been debased and devalued by capitalists (Kanyenze 2018, 120).

The foregoing conclusion that I, thus, advance in this chapter is that whilst the Fast-Track Land Reform Programme of the 2000s and the Climate Change perspectives including the social crises argument advanced elsewhere (Chiba and Thebe 2021) are correct assertions in explaining the Zimbabwean Crisis, here we shift the angle of analysis by arguing that the lack of democratic space, especially through the decline of civil society groups (trade unions in particular) from the 1990s owing to the competitive nature of Zimbabwe's economy bred a whole array of multiple crises that is currently rocking the Zimbabwean Society. This implies that democracy should be the first important step if an egalitarian and developed society is to be realised. The Sustainable Livelihoods Framework, thus, allowed me to ascertain how the macro-environment, through the generation of unemployment (Transforming structures and processes), have affected the functioning of smallholder farmers' five capitals (social, human, natural, financial and physical) and hence their agricultural production processes and general livelihoods.

Conclusion

Wage employment is an important livelihood strategy with a long-pedigree in Zimbabwe. It contributes to the success of the smallholder agricultural enterprise, acts as a source of food security and reduces poverty among households. However, the formal unemployment crisis that has negatively affected the Zimbabwean economy since the 1990s has brought typical rudiments of counter-productivity to smallholder agriculture. The crisis itself has led to the periodic shedding of assets or even a failure to update these.

There is, thus, a need, on the part of the government, to create employment, may be through influencing both fiscal and monetary policies to broaden the industrial base. This may also be done through the tight management of skills training by managing schools, colleges and universities in a responsible manner so that training of people at least cancels out with available employment opportunities. This will then address other new crises which are surfacing in late capitalist economies including that of social protection in old age.

References

Achy, L .2011. Tunisia's Economic Challenges. *The Carnegie Papers.*

Alexander J, McGregor J & Ranger T. 2000. *Violence and Memory: One Hundred Years in the Dark Years in the 'Dark Forests' of Matabeleland.* Oxford: James Currey.

Arrighi, G .1970. 'Labour supplies in historical perspective: A study of the proletarianization of the African peasantry in Rhodesia'. *Journal of Development Studies VI*: 198–233.

Ashley, C & Carney, D.1999. *Sustainable livelihoods: Lessons from early experience* (Vol. 7, No. 1). London: Department for International Development, USA.

Bernstein, H.2004. 'Changing before our very eyes': Agrarian questions and the politics of land in capitalism today. *Journal of Agrarian Change*, *4*(1-2): 190-225.

Bird, K & Shepherd, A.2003. Chronic Poverty in Semi-Arid Zimbabwe. *Overseas Development Institute*, CPRC Working Paper No. 18: 1 – 59.

Brett, E .2005. 'From corporatism to liberalization in Zimbabwe: Economic policy regimes and political crisis, 1980–97'. *International Political Science Review26* (1): 91–106.

Bush, R & Cliffe, L.1984. 'Agrarian policy in migrant labour societies: Reform or transformation in Zimbabwe?' *Review of African Political Economy* 29: 77–94.

Bryceson, D.2000. Rural Africa at the crossroads: Livelihood practices and policies. London: Overseas Development Institute.

Chirisa, I.2013. Social Protection amid Increasing Instability in Zimbabwe: Scope, Institutions and Policy Options: In Stephen Devereux Melese Getu. *Informal and Formal Social Protection Systems in Sub-Saharan Africa* (Organisation for Social Science Research in Eastern and Southern Africa (OSSREA) Fountain Publishers: 121-158.

Chiba, M. and Thebe, V., 2022. Changing Households Social Dynamics and Agriculture Crisis in Shamva District, Zimbabwe. *Journal of Asian and African Studies*, p.00219096221076161.

Chiumbu, S & Musemwa, M.2012. Introduction: Perspectives of the Zimbabwean Crises. In *Crisis! What Crisis? The Multiple Dimensions of the Zimbabwean Crisis*, edited by Chiumbu S & Musemwa M. Human Sciences Research Council, Cape Town, South Africa.

Clover, J.2003. 'Food security in sub-Saharan Africa', *African Security Studies, 12*(1): 5-15.

Cousins, B, Weiner, D, and Amin, N.1992. 'Social differentiation in the communal lands of Zimbabwe'. *Review of African Political Economy, 19*(53): 5-24.

Crush J, Chikanda A, & Tawodzera, G. 2015. 'The third wave: Mixed migration from Zimbabwe to South Africa.' *Canadian Journal of African Studies/Revue canadienne des études africaines* 49(2): 363–382.

Denzin, N.K & Lincoln, Y.S eds. 2011. The Sage handbook of qualitative research. *Sage, Development 4(1):1-14.*

Dorward A, Anderson S, Clark S, Keane B, & Moguel, J.2001. Asset Functions and Livelihood Strategies: A Framework for Pro-Poor Analysis, Policy and Practice. *Contributed Paper to EAAE Seminar on Livelihoods and Rural Poverty, September 2001.*

Duggan, W.R. 1980. 'The Native Land Husbandry Act of 1951 and the rural African middle class of southern Rhodesia'. *African Affairs, 79*(315):227-239.

Ellis, F & Biggs, S. 2001. 'Evolving Themes in Rural Development 1950s-2000s.' *Development Policy Review 19 (4)*: 437-448

Fink, G, Kelsey, B, Masiye J.F, Jack B.K, & Masiye, F .2014. *Seasonal credit constraints and agricultural labor supply: Evidence from Zambia* (NBER Working Paper Series No. 20218).

Fleşer, A & Dobre-Baron, O.2010. Economic and Social Implications of Unemployment and Opportunities of Diminishing it. Tome VIII, *Fascicule 3*:72-77.

Hayami, Y. 1996. 'The Peasant in Economic Organization'. *American Journal of Agricultural Economics. Vol 78 (5):* 1157-1167.

Hoddinott, J.2006. 'Shocks and their consequences across and within households in rural Zimbabwe.' *The Journal of Development Studies, 42*(2), pp.301-321.

Holzmann, R.2001. 'Social Risk Management: A New Conceptual Framework for Social Protection, and Beyond'. *International Tax and Public Finance, 8*, 529–556. https://tradingeconomics.com (accessed 8 March 2022).

Kanyenze, G. 2018. Economic Crisis, Structural Change and the Devaluation of Labour: In: Sachikonye, L; Raftopoulos, B & Kanyenze, G., *Building from the Rubble: The Labour Movement in Zimbabwe since 2000.* Friedrich-Ebert Stiftung-Weaver Press, Harare

Khan, S.N. 2014. 'Qualitative research method: Grounded theory.' *International Journal of Business and Management, 9*(11):224-233.

Lundahl M, McCarthy C, & Petersson L.2003. *In the Shadow of South Africa: Lesotho's Economic Future* Ashgate Aldershot.

Malepe, L.2014. From Youth Inequality to Youth Participation in Socio-Economic Development: In: Stiftung FE, Youth, *Regioness and SADC Action for and by Youth to build a brighter future.*

Mandelman, F.S & Montes-Rojas G.V.2009. 'Is self-employment and micro entrepreneurship a desired outcome?' *World Development*, 37, 1914-1925.

Mano, R, Isaacson B, & Philippe, D.2003. *Identifying Policy Determinants of Food Security Response and Recovery in the SADC Region: The Case of the 2002 Food Emergency* FANRPPAN Policy Paper.

Marquette, C.1997. Current poverty, structural adjustment and drought in Zimbabwe. *World Development* 25(7): 1141–1149.

Matondi, P.2012. *Zimbabwe's Fast Track Land Reform.* The Nordic Africa Institute, Zed Books, London/New York.

McMichael, P.2006. Peasant prospects in the neoliberal age. *New Political Economy, 11 (3)*:407-418.

Mhazo, T. and Thebe, V., 2021. 'Hustling Out of Unemployment': Livelihood Responses of Unemployed Young Graduates in the City of Bulawayo, Zimbabwe. *Journal of Asian and African Studies*, *56*(3), pp.628-642.

Mosley, P.1983. *The settler economies: studies in the economic history of Kenya and Southern Rhodesia 1900–1963.* Cambridge: Cambridge University Press.

Moyo, S.2005. The Land Question and the Peasantry in Southern Africa. Latin American Council of Social Sciences: 145-163.

Moyo, S.2007. The land question in southern Africa: a comparative review. The Land Question: In: Ntsebeza L & Hall, R., *The Land Question in South Africa: The Challenge of Transformation and Redistribution.* HSRC Press.

Moyo, S.2011. Changing Agrarian Relations after redistributive Land Reform in Zimbabwe. *The Journal of Peasant Studies* 38(5):939-966.

Moyo, S & Yeros, P.2007. The radicalised state: Zimbabwe's interrupted revolution. *Review of African Political Economy*, *34*(111):103-121.

Mtetwa, R.M.G.1976. *The Political and Economic History of the Duma People of South-Eastern Rhodesia from the Early Eighteenth Century to 1945.* An Unpublished doctoral Thesis, University of Rhodesia.

Mtimtema, Z. 2018. Political, Judicial and Legislature Responses to Labour and the Changing Regime of Industrial Relations: In: Sachikonye, L; Raftopoulos, B & Kanyenze, G., *Building from the Rubble: The Labour Movement in Zimbabwe since 2000.* Friedrich-Ebert Stiftung-Weaver Press, Harare.

Murray, C.1981. *Families Divided: The Impact of Migrant Labour in Lesotho. Cambridge.* African Studies Series.

Page, S.L & Page, H.E.1992. Western hegemony over African agriculture in Southern Rhodesia and its continuing threat to food security in independent Zimbabwe. *Agriculture and Human Values*, *8*(4), pp.3-18.

Potts, D.2000. Urban unemployment and migrants in Africa: Evidence from Harare 1985–1994. *Development and Change 31*: 879–910.

Potts, D & Mutambirwa, C.1990. Rural-urban linkages in contemporary Harare: why migrants need their land. *Journal of Southern African Studies*, *16*(4): 677-698.

Prescott, J.R.V.1961. 'Overpopulation and overstocking in native areas of Matabeleland', *Geographical Journal 127*, 2:212 –225.

Puentes, E, Contreras, D & Sanhueza, C.2007. Self-employment in Chile, long run trends and education and age structures changes. *Estudios de Economia, 34*, 203-247.

Raftopoulos, B., 2006. The Zimbabwean crisis and the challenges for the left. *Journal of Southern African Studies*, *32*(2), pp.203-219.

Richardson, C.J.2005. The loss of property rights and the collapse of Zimbabwe. *Cato Journal (25):*541-565.

Richardson, C.J.2007. How much did droughts matter? Linking rainfall and GDP growth in Zimbabwe. *African Affairs*, 106(424): 463-478.

Scoones, I.1998. Sustainable Rural Livelihoods: A Framework for Analysis. *IDS Working Paper 72.*

Stoneman, C.1991. Jobs or Markets: Lessons from Zimbabwe. Paper presented to Centre for African Studies, University of Cape Town, October 1991.

Symon, G & Cassell, C.2004. *Promoting New Research Practices in Organizational Research. Essential guide to qualitative methods in organizational research.* Sage Publications.

Tadross, M.A, Hewitson B.C & Usman M.T.2005. The Interannual variability of the onset of the maize growing season over South Africa and Zimbabwe. *Journal of Climate, 18*(16):3356-3372.

Thebe, V. 2012. 'New Realities' and tenure reforms: Land-use in worker peasant communities of south-western Zimbabwe (1940s–2006). Journal of Contemporary African Studies, 30(1), 99–117.

Thebe, V., 2014. New Social Formations, Livelihood Transition, and Food Insecurity in Worker–Peasant Communities. *Journal of Developing Societies*, *30*(1), pp.45-67.

Van der Ploeg, J.D.2020. From bio-medical to politico-economic crisis: the food system in times of Covid-19. *The Journal of Peasant Studies:* 1-30.

Van Hoyweghen K, Van Den Broeck G & Maertens, M. 2018. Understanding the importance of wage employment for rural development: Evidence from Senegal. Division of Bio-economics Department of Earth and Environmental Sciences. Working Paper 03.

Waeterloos, E & Rutherford, B.2003. 'Land reform in Zimbabwe: Challenges and opportunities for poverty reduction among commercial farmworkers.' *World Development 32 (3)*: 537–553.

Worby, L.E.2001. 'A redivided land? New agrarian conflicts and questions in Zimbabwe'. *Journal of Agrarian Change 1* (4): 475–509.

World Bank. 2008. *World Development Report 2008: Agriculture for Development* (Washington, DC).

World Health Organisation.2020. Coronavirus disease (COVID-19- 2019) situation reports. Accessed June 30, 2020. https://www.who.int/emergencies/diseases/novel-coronavirus-2019/situat ion-reports/ www.bbc.com/news/business-42116932) accessed 8 March 2022

Xu, M., David, J.M. and Kim, S.H., 2018. The fourth industrial revolution: Opportunities and challenges. *International journal of financial research*, *9*(2), pp.90-95.

Chapter 3

Northern Development NGOs as Special Interest Groups: How the Smallholder 'Idea' Promotes Agricultural Protectionism

Andrea Franc[1,*]
and Stefan Mann[2,†]
[1]University of Lucerne, Switzerland,
[2]Agroscope, Ettenhausen, Switzerland

Abstract

Based on an application of Mancur Olson's collective action approach, this paper draws attention to the role of Northern development NGOs in advancing agricultural protectionism. It challenges the current dogma of mainstream economics that the farm lobby is solely responsible for agricultural protectionism in the North. It can be shown that Northern development NGOs have been demanding agricultural protectionism since the 1970s. The narrative on fair (i.e., sustainable) trade has changed over time, following the professionalisation of the sector. In the years after the United Nations Conference on Trade and Development of 1964, spontaneous activist groups demanded the industrialisation of developing countries. By the time professional NGOs and their private labels for food products emerged in the 1990s, the narrative had narrowed on smallholder farming and a restricted basket of tropical food commodities. The promotion of family farms in the North and poor peasants in the South is a promising way for the maximization of donations but builds political momentum for a system of border protections that,

* Corresponding Author's E-mail: andrea.franc@doz.unilu.ch.
† Corresponding Author's E-mail: stefan.mann@agroscope.admin.ch.

In: Agricultural Policy
Editor: Youssef Lee
ISBN: 979-8-89113-720-2

paradoxically, prevents peasants to benefit from trade gains. When formulating sustainability policies, governments should therefore not adopt Northern development NGOs criteria.

Keywords: PPMs, fair trade, NGOs, development, sustainability, special interest groups

1. Introduction

The impact of Mancur Olson's "Logic of Collective Action" (1965) with its more than 50 000 citations for research in political economy can hardly be overestimated (Olson 1965). To date, the idea that engagement in interest groups can maximize the utility of group members continues to find application in realms ranging from agri-environmental land use conflicts in Africa (Willy and Ngare 2021) to lobbying in the European Union (Bruycker, Berkhout, and Hanegraaff 2019). Less attention, however, has been devoted to a contribution by Johnson and Prakash (2007) who argued: "By employing a collective action perspective (…), NGO scholars will be able to develop explanations about NGO origin, structure, and strategy that have superior explanatory power and are generalizable across NGO's (Johnson and Prakash 2007, 221)."

This research dwells on this approach to understand the dynamics of Northern development nongovernmental organisations (NGOs) in their attitude towards agricultural protectionism. It challenges the current dogma of mainstream economics that farm lobbies are solely responsible for the agricultural fortress of Europe (Dümmler and Anthamatten 2020; Krugman, Obstfeld, and Melitz 2015). This paper argues that not only the farm lobby, but more so the narrative (i.e., the operationalisation of sustainability as promoted by Northern development NGOs) has become key to understanding agricultural protectionism in rich industrial nations.

For this purpose, the paper proceeds as follows: In Section 2, the claim about the NGOs' narrative is developed. The method to test the claim is described in Section 3. Section 4 presents results, and Section 5 concludes.

2. The Theory of Special Interest GROUPS and How to Expand It

Economics textbooks teach that special interest groups (SIGs) distort the democratic process by lobbying for their members (Krugman, Obstfeld, and Melitz 2015, 280–82). This theory on SIGs is still based on Olson's (1965) "Collective Action" and Grossman and Helpman's (1994) "Protection for Sale" (Grossman and Helpman 1994). Olson's collective action theory has repeatedly been applied to argue that agricultural protectionism is explained by the incentive for small groups to organise and lobby in their members' interests (Thies 2015; Malang and Holzinger 2020). This is certainly a suitable approach: compared with non-organised, heterogeneous groups, SIGs are organised, and this allows them to better protect their interests. As a perfect example of Olson's collective action theory, an organised farmers' union can effectively protect high beet sugar prices and prevent the import of cheap cane sugar. Non-organised and heterogeneous consumers not only have trouble lobbying to lower sugar prices but also might not even notice the effects of protection, as the price of sugar would be insignificant to consumers even, when artificially inflated through protection.

Grossman and Helpman's (1994) theory posits that SIGs buy protection for their members through 'buying' politicians through campaign donations. However, this empirical observation reveals interesting differences in the impact of SIGs, which are only quantifiable to a certain degree. As Frey (1985) noted, SIGs in economically irrelevant sectors sometimes successfully obtain protection, rather than those with the highest financial means winning the most protection, such as the pharmaceutical or financial industry lobbies (Frey 1985). Sometimes other 'capital', such as important voter areas, might explain an SIG's success. Rodrik (2014) expanded the Grossman–Helpman theory by introducing the concept of an 'idea' (e.g., a 'worldview') that might trump special interests, thereby pushing the boundaries of mainstream economics and inviting qualitative approaches (Rodrik 2014).

Which impact has the conceptualisation of fair trade, and later sustainability, advocated by Northern development NGOs? To answer this question, it is important to discuss three hypotheses:

1. Political economists have become sympathetic to the idea that NGOs tend to maximize donations (Heyes and Martin 2015; Kuhnert 2018).

Collective action theory therefore can also be twisted to fit a small group of *employees* of an NGO as beneficiaries of donations. Development NGOs provide no exception to this rule.

2. Donations can be maximized by allegedly defending the interest of the weakest part of society. Conservative and right-wing voters and politicians are intuitively understood to support the protection of domestic family farms as part of their 'worldview' as the backbone of national identity (Franc 2018). Urban, progressive voters demand solidarity with peasants worldwide. Behind this worldview is the idea of local, organic food production, set forth in in the 1970s bestsellers "Small is Beautiful" (Schumacher, 1973), "How the Other Half Dies" (George, 1976), and "Food First" (Schumacher 1973; George 1976; Lappé, Fowler, and Collins 1977).
3. With this focus, it is always useful in an NGO's communication to dismiss 'big business' and globalisation. As international trade necessarily involves larger organizations, this results in a protectionist narrative that prevents open trade, even if the latter eventually would benefit peasants in the global south.

Donor-dependent development NGOs are SIGs, but a particular type, because *officially*, they do not promote the interests of their members (or non-member donors) let alone their employees, but of the people *on whose behalf* donations are made: the poor in the global south. This makes it tricky to understand development NGOs as SIGs, especially in national agricultural protectionism. Grossman and Helpman's theory posits that SIGs make campaign contributions to politicians. As can be empirically observed, NGOs do 'contribute' to the political process, albeit in different forms. Their currency is 'moral' capital, which they source from their well-tended grassroots, dating back to 1968 (Buchanan 2002). Instead of financing a campaign, they endorse certain candidates with their 'moral' capital by recommending them to their members or donors.[1] Likewise, NGOs can use their 'moral' capital to destroy a politician's reputation, the threat of which alone might be considered a significant 'contribution' to the political process.

Grossman and Helpman (1994) assume that all SIGs strive to maximise their members' welfare. When examining development NGOs in real-world

[1] See, for instance, the endorsement list of a group of NGOs, such as Greenpeace or WWF https://ecorating.ch/de (21.6.2023)

settings, the fundamental divergence between *professed* and *pursued* goals makes Grossman and Helpman's SIG theory difficult to apply. While both the professed and pursued goals of a union is to save their members jobs, or in the case of a farmers' union is to increase member subsidies, the *professed* goal of Oxfam is to end poverty in the South, while its *pursued* goal must necessarily be to ensure if not increase the steady influx of donations to maintain an organisation in the North. Donors in developed countries have 'preferences, world views and ideas' about the world and world poverty (Rodrik 2014). Therefore, to maximise donations, NGOs must necessarily not actually end world poverty but cater to donors' 'idea' of how to end world poverty. World poverty is alleviated by creating a sufficient number of steady jobs in productive sectors; however, donors might object to the idea (promoted by Duflo and Banerjee) of multinational corporations' sweatshops and call centres in Latin America and India (Banerjee and Duflo 2011, 297). Donors might prefer organic smallholder farming, despite the fact that it encourages smallholders "to stay doing what they are doing - producing coffee". To continue quoting Collier, donors do not like to be reminded that "[a] key economic problem for the bottom billion is that producers have not diversified out of a narrow range of primary commodities. [...] They get charity as long as they stay producing the crops that have locked them into poverty" (Collier 2007, 163). Donor-dependent Northern NGOs do not propose that smallholders plainly exit agriculture and enter the industrial or service sector.

The impact of the narrative of development NGOs does lead lawmakers to frame laws allowing the World Trade Organisation (WTO) to conform to product differentiation (processes and production methods, PPM), thereby actively encouraging imports from smallholders in developing countries (Baumgartner and Bürgi Bonanomi 2021). However, NGOs have pushed the development and sustainability narrative to narrowly focus on smallholder agriculture (Bürgi Bonanomi 2015). Instead of crediting Northern development NGOs with feeble initiatives that encourage a tiny basket of agricultural food imports by certified smallholder producers, the same NGOs should be held accountable for promoting excessive agricultural protectionism since the 1970s (Anderson 2015; Jordan and Maloney 1997; Moser 1994). As long as development NGOs are considered to be altruistic grassroots organisations representing the poor in the global south, lawmakers will simply translate private label definitions into WTO conform PPM laws (Mazidi, Simon, et al.).

3. Qualitative Historical Research with NGO Records

The method used here is qualitative historical archival research. The qualitative data comprises records of NGOs and civil society movements that are actors in agricultural policy and promote labels for fairly traded or sustainably produced food. Records are fully available for Public Eye, the NGO most active in agricultural policy in Switzerland.[2] The records of Oxfam, the best-known NGO internationally for influencing agricultural policy, are currently classified, but have been consulted as well to crosscheck the case study on Public Eye.[3] Prominent actors from both Public Eye and Oxfam have repeatedly collaborated over the last few decades, most notably during the World Food Crisis of the mid-1970s, when a new paradigm took shape on the importance of diversified, smallholder farming systems worldwide (Gerlach 2015). The data are well-suited to explaining the global impact of the smallholder narrative using a case study on Switzerland.

Although the Swiss NGO Public Eye is small on an international level, their early activists are historically famous for the lawsuit by Swiss multinational company Nestlé filed in response to "The Baby Killer", a 1974 pamphlet condemning infant formula, written by Mike Muller and published by the British NGO Christian Aid. This was when multinational corporations became targets of NGO pamphlets, and politicians associated with those corporations began to fear for their reputations. Nestlé only reacted when Swiss activists translated the brochure into German and made it known in Switzerland. One activist Nestlé accused in the lawsuit was economist Rudolf Strahm, then-secretary of the NGO Berne Declaration (now Public Eye) (Sasson 2016). This incident exemplifies that the records of the small Swiss NGO are qualitative data of global relevance and illustrate the NGO's impact as an SIG. Furthermore, the case study is embedded in the findings of a larger research project, the Database of Archives of Non-Government Organisations (DANGO), conducted in the United Kingdom over several years (Hilton 2013).

---

[2] See Swiss Social Archives in Zurich, signature EvB, Ar 430.

[3] See Special Collections at the Bodleian Library, Oxford. Further records are kept as pamphlets at the Social Science Library, Oxford. Other British NGOs, such as Christian Aid, gave their records to the Archives of the School of Oriental and African Studies, London, where they can be consulted.

4. Professionalisation of Third World Advocacy Groups: Changing the Narrative on 'Fair Trade'

While the trade liberalisation momentum of the Kennedy Round of negotiations fit the revolutionary spirit of grassroots activists of the 1960s, the prospect of structural upheaval does not appeal to either donors or employees of professional NGOs (McAdam 2021).[4] In 1968, the Swiss Berne Declaration stated that "Switzerland had to abandon privileges" and "structural change is inevitable".[5] However, in the 1980s, employees of newly established NGOs quickly reverted to preserving local smallholder farming, effectively turning development NGOs into protectionist SIGs and political partners of farmers' unions.

Only in recent years has the term 'sustainability' overtaken the terms 'fair trade' and 'organic farming' in the narrative of development NGOs. The term sustainability has allowed for the conflation of both fair trade and organic farming. Private labels originally developed within the narrative on fair trade and organic farming and only later adopted the term sustainability. In the 1980s, the term fair trade was absent, and political terms such as solidarity, food security, and food sovereignty dominated the discourse. In the 1960s and early 1970s, agriculture, let alone smallholder farming, had not yet entered the narrative field. The movement was in the hands of trained economists or lawyers who spoke and wrote in their academic jargon, using the term fair trade in its initial sense of the most favoured nation principle of the General Agreement on Tariffs and Trade (GATT). Regarding this early narrative, Oxfam activist Maggie Black later stated:

> "A hundred or so committed development action groups beavered away, trying to disentangle growth rates from commodity agreements, unearth the mysteries of ODAs and GATTs, unhook multilaterals from intergovernmentals, and work out where the poor fitted in (Black 1992, 159)."

Owing to the profound change in concept and narrative that has occurred since, private fair-trade labels have been thought to originate in the 1990s (Nicholls and Opal 2005). In the 1990s, NGOs separated the term fair trade

[4] Mark McAdam has noted that the 'idea' of trade liberalisation as national interest prompted the Kennedy Round, see (McAdam 2021).
[5] https://www.publiceye.ch/de/was-wir-tun/50-jahre-public-eye (22.6.2023).

from its original use in the GATT and used it as a new label for smallholder-produced tropical commodities. Dutch historian Peter van Dam has only recently traced the commercial fair-trade movement of the 1990s back to 1968, which is the year activists in the Netherlands first used a tropical commodity, cane sugar, as a symbol for the movement (van Dam 2016). Van Dam's findings on the Netherlands also show that, originally, market access for cane sugar, which suffered from a competitive disadvantage vis-à-vis European subsidised beet sugar, was the first commodity for which the activists petitioned. These activists used cane sugar to exemplify the exploitation of the Third World, expose an unfair trading system, and support the demands of the United Nations Conference on Trade and Development (UNCTAD). In particular, the Dutch activists were early to denounce the Common Agricultural Policy of the European Economic Community.

The original UNCTAD and GATT jargon fair-trade concept, which encompasses the economy as a whole, which includes all three sectors (i.e., agriculture, industry, and service) and promotes industrialisation of developing countries including those in Africa, becomes particularly evident in 1973, with the Tanzanian instant-coffee campaign in Switzerland, the United Kingdom, the Netherlands, and Germany. At the time, activists considered industrial products from an African factory to be fair, and instant coffee produced in a Tanzanian factory exemplified the need to open European markets to Southern industrial goods and apply the General System of Preferences. Another such example is the 1973 Alu-Schok campaign in Germany, which also focused on local industrial processing of tropical commodities: bauxite and cocoa (i.e., aluminium and chocolate) (Möckel 2019). The DANGO project revealed the professionalisation of the NGO sector that surprisingly flourished during the neoliberal phase of the 1980s and 1990s (Hilton 2013, 191–97). When the responsibilities of the British state contracted and people lost faith in political parties, NGOs took over and provided social services and a new value system. In Switzerland, the privatisation of public sector services was less pronounced compared with in Britain, but the loss of interest in traditional politics is equally observable. Voter participation dropped from almost 70 per cent in 1970 to slightly over 40 per cent in 1990, where it has remained until today (Neidhart 2002-2014). The commercial fair-trade trademark was created in the context of this privatisation of politics in Western Europe. Furthermore, the church lost its predominant role in society. Anderson examined the role of traditional churches regarding fair trade and found that churches secularised their narrative and actively shaped and supported the introduction of private labels in the United Kingdom (Anderson 2019). A similar observation can be made for Switzerland, as the official development NGOs of the Catholic and Protestant churches established the private fair-trade label.[6]

The about face in the fair-trade narrative becomes apparent when reading pamphlets from the 1960s. The concept of fair trade was originally based on academic economic literature of the time (Haslemere group [1969]). In Britain, a group comprising journalists, academics, and employees of the main British Third World charities met in Haslemere, Surrey, in January 1968 and published the Haslemere Declaration (Haslemere group 1968, 3). The Haslemere Group's reading guide, published in 1969, details what the members had read and what the many local action groups that arose after the publication of the declaration (and to whom Maggie Black refers) would read (Black 1992, 159; Haslemere group [1969]). Although it does not come as a surprise that the viewpoints of the Haslemere activists were informed by Marxist and Keynesian literature, a closer look at the guide's content offers a surprising insight into the Third World activists' critical attitude towards the unification of Europe – a feature that is completely absent in the NGO narrative today. Along this line, they recommend Michael Barratt Brown's (1963) "After Imperialism", which mainly examines British overseas investments in the 1950s and, based on this examination, is critical of the Common Market and the effect of an eventual British entry and consequent market access of the Commonwealth nations (Brown 1963, 372–80). According to the activists, although industrialisation must be achieved through self-reliant state policies in developing countries, market access to Europe was threatened by the ongoing unification process (Brown 1963, 379).

As Haslemere Group economist Jonathan Power wrote in 1969, the creation of Europe as a protectionist fortress was seen "another nail in the coffin of the Third World" (Jonathan Power 1969, 3). In another clue to the understanding of fair or unfair trade in the 1960s, the reading guide suggests books by Southern politicians and intellectuals, such as Kwame Nkrumah, Mahatma K. Ghandi, and K. M. Panikkar. The figure of the Southern intellectual is prevalent, while the Southern 'smallholder' as either the victim or recipient of charity is absent.[7]

---

[6] See https://www.fairtrademaxhavelaar.ch/ueber-uns/in-der-schweiz/geschichte (22.6.2023)

[7] Furthermore, the economic section of the reading guide suggests titles that would be expected, such as those by André Gunder Frank, Lenin, Ernest Mandel, and Gunnar Myrdal.

At the same time in Switzerland, a group of Protestant academics launched a declaration similar to the "Haslemere Declaration". They drafted the "Berne Declaration" in March 1968, signed by 1000 prominent intellectuals in Switzerland, and delivered it to the Swiss government in January 1969. Over the years to come, the professional NGO Berne Declaration, renamed Public Eye in 2016, would slowly emerge from this spontaneous and honorary grassroots movement. The group of Protestant theologians involved in the draft knew each other as delegates or employees of the World Council of Churches in Geneva, which at the time was a 'hot spot' of Latin American Liberation theology and dependency theory (Dols and Ziemann 2015). Through the World Council of Churches and the first UNCTAD conference in Geneva in 1964, as their Christian duty, this group redefined the North–South trade issues of market access and industrialisation, known as fair trade in the narrative of time. This informal group plunged into North–South economics and gathered regularly to learn more about the Prebisch–Singer thesis, import substitution investment, and infant industry protection.

In April 1967, they invited Swiss lawyer Christoph Eckenstein to present a paper on UNCTAD's demands, and he has worked closely with them since (Gessler 2008). Eckenstein is a seasoned development consultant and known as the architect of the legal framework of the preferential tariff system, the General System of Preferences (Eckenstein and Gurtner 1977, 13). He began working for the United Nations Economic Commission on Latin America and the Caribbean in collaboration with Prebisch in Latin America in 1963 and helped prepare the UNCTAD of 1964 in Geneva. When Prebisch became the Secretary-General of UNCTAD, Eckenstein became Special Advisor on Trade at UNCTAD in Geneva. In 1970, Prebisch spoke at a conference they organised in the Swiss capital of Berne to discuss further actions after the "Berne Declaration" (Traber and Schmocker 1971, 24).

The key difference between the British Haslemere Group, which comprised employees of charities, and the Swiss group, which consisted of slightly older theologians, some of them professors, lay in the way that they envisaged industrialisation in developing countries. The Haslemere Group was highly critical of private investments and supported socialist-state development of industry. The Swiss group did not in any way promote socialist-state intervention, but very clearly demanded that Swiss companies invest overseas instead of merely extracting commodities (Erklärung von Bern 1970b). However, both groups promoted import substitution investment and infant industry protection (i.e., industrialisation and economic growth of developing countries). They primarily made it their mission to raise awareness for the need for market access for these countries' products. Market access and industrialisation are North–South issues and can technically be achieved under either capitalism or socialism. It was their focus on these two main North–South issues that related these two groups intellectually (Erklärung von Bern 1968).

In a first step towards becoming the professional NGO it is today, in 1969, the Berne Declaration[8] employed a secretary, Anne-Marie Holenstein. In the same year, a group of Swiss students, among them economics student and future co-secretary Rudolf Strahm, visited the Haslemere Congress "On Race and Poverty" in Manchester. They brought a number of copies of the "Haslemere Declaration" back to Switzerland and distributed them to activists (Erklärung von Bern 1969). In spring 1970, Holenstein travelled to Egmond aan Zee in the Netherlands and participated in a meeting of European activists to discuss a coordinated action regarding barriers to trade for cane sugar from developing countries (Erklärung von Bern 1970a) (Holenstein, Renschler, and Strahm 2008, 50). In retrospect, this was the crucial meeting for the start of the commercial fair trade movement, which would form a network of world shops and later establish labels in Europe (van Dam 2019). Clearly, the original academic concept of fair trade that stood for market access for all existing and potential Southern products (not only tropical commodities) led straight to a research-based critique of supranational organisations, such as the European Economic Community or the GATT, that stood for the reinforcement of the 'privileges' and the trade barriers of industrial countries.

[8] Now written without quotation marks to designate the NGO and not the declaration.

Etiquette Kaffeeaktion 1973

Warum Kaffee hergestellt in Tansania?

Wenn Sie Pulverkaffee kaufen, der in der Schweiz verarbeitet worden ist, bezahlen Sie durchschnittlich Fr. 50.— für 1 Kilogramm. Davon gehen aber nur Fr. 10.— ins Ursprungsland. Dabei ist der Kaffeebaum schwierig zu pflegen, beansprucht den Boden stark und bringt während der ersten fünf Jahre praktisch keine Frucht.

Um 1 kg Pulverkaffee zu erhalten, muss der Kaffeebauer 12 000 Kaffeebohnen (d. h. die Ernte von 6 Kaffeebäumen) pflücken, waschen, sortieren und schälen – das alles für 10 Franken. Das Einkommen der Kaffeebauern ist dementsprechend gering.
Die Entwicklungsländer sollten den Kaffee selber verarbeiten können. Die dazu nötige Industrialisierung aber wird durch ihren Kapitalmangel und durch unser Zollsystem behindert.

Der UJAMAA-Kaffee ist nun in Tansania selber verarbeitet worden, d. h. im Land, wo die Bohnen gepflückt worden sind. Das bedeutet:

- Tansania bekommt nicht Fr. 10.—, sondern Fr. 16.— pro kg Pulverkaffee.
- An der Verarbeitung gewinnen tansanische Arbeiter und nicht irgend ein Grosskonzern.
- Es ergeben sich damit neue Arbeitsplätze für Tansanier.
- Tansania erzielt stabile Preise für sein Produkt, denn Pulverkaffee kann gelagert werden, Kaffeebohnen jedoch nicht; sie müssen jeweils [illegible]ofort abgestossen werden, oft [illegible]de dann, wenn der Preis extre[illegible] niedrig ist.

UJAMAA KAFFEE

Der Kaffee mit dem Aroma der Selbstbefreiung

Reiner Kaffeeextrakt
Hergestellt in TANSANIA
100 gr. ergeben ca. 66 Tassen

Prei[illegible]r. 3.60

Warum Kaffee UJAMAA?

Dieser Kaffee ist nicht das bittere Produkt brasilianischer Hungerlöhne und Foltern.

Er hat nicht den bitteren Beigeschmack vom Blut jener Angolesen, die unter portugiesischem Terror schuften müssen.

Und er wurde nicht durch unsere Industrie verteuert.

Er kommt nämlich aus Tansania, wo er geschält, getrocknet, gebrannt und gemahlen wurde. Tansania ist jenes afrikanische Land, wo die Armen angefangen haben, sich selber zu befreien. Und zwar aus eigener Kraft.

Das Mittel zu dieser Selbstbefreiung heisst UJAMAA:
- gemeinsam planen
- gemeinsam arbeiten
- gemeinsam verdienen
- gemeinsam leben

Figure 1. Stick-on label for Tanzanian instant coffee made in a factory in Bukoba (AG3W 1973).

While the Haslemere Group published several pamphlets on the original academic concept of fair trade in Britain, in Switzerland, Strahm finished his master's thesis in economics, in which he tested the Prebisch–Singer thesis in the Swiss context (Strahm). In 1973, Strahm managed to import a shipment of instant coffee from a Tanzanian factory in Bukoba (Strahm and Hug 1975). Across Switzerland, he trained volunteers in so-called 'selling workshops', who then sold the instant coffee from stands in high streets. They printed the message that they wanted to see conveyed on all such labels and posters: "with industrial value added, Tanzania earns more" and "our import barriers hinder imports from developing countries (Figure 1) (AG3W 1973)".

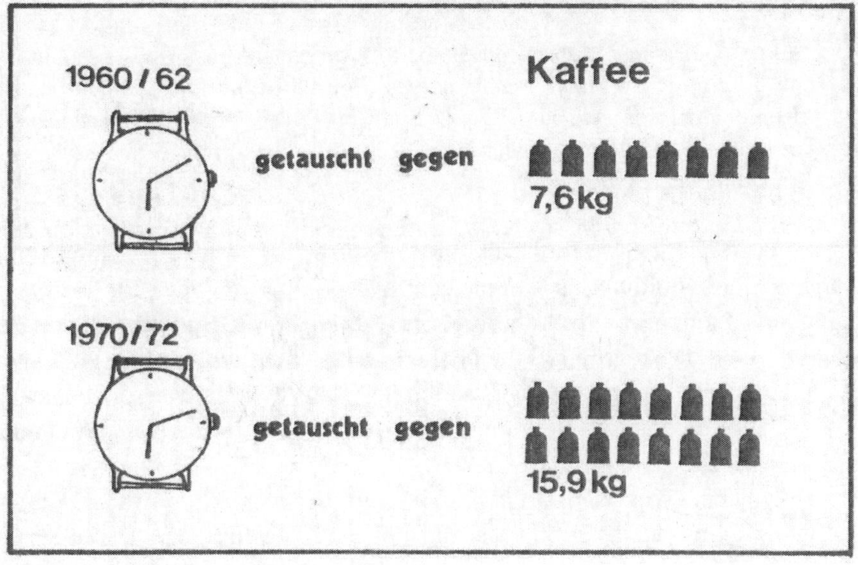

Figure 2. Graph explaining the Prebisch–Singer thesis to volunteers (Strahm and Hug 1975).

In addition, the image of the smallholder farmer was practically absent in the campaign, and the Tanzanian instant coffee was not sold with a charity premium but explicitly cheaper than Nescafé (Erklärung von Bern [ca. 1975]). After the first instant-coffee campaign was launched, Strahm worked as a consultant at the UNCTAD in Geneva and travelled to Tanzania (Holenstein, Renschler, and Strahm 2008, 113–66). Afterwards, in 1974, he joined Holenstein as co-secretary. Under the auspices of the new NGO Berne

Declaration, they repeated the instant-coffee campaign twice, in 1974 and 1975 (Strahm and Hug 1975). For the last sale in 1975, Strahm produced official training materials for volunteers who sold the instant coffee from high street stands (Figure 2).

This material illustrates again how the volunteers were trained in the original, academic concept of fair trade. First, a simple graph explained the Prebisch–Singer thesis based on the number of bags of coffee beans that are needed in exchange for a Swiss watch. Then, the volunteers learned about unfair barriers to trade that prevented market access and the need for developing countries to industrialise. One volunteer summarised this as follows:

> "The coffee campaign shall convey the following message: (…) We should not only import primary commodities from the developing countries (deterioration of prices) but also industrially processed goods (e.g., Tanzania earns 60% more by selling instant coffee than by exporting coffee beans) (Erklärung von Bern 1974)".

After 1977, the Berne Declaration assumed its form as an intellectual NGO, which would act as a think tank and produce an 'idea': information, pamphlets, campaigns, and conferences. Strahm left the administration, the authors of the declaration left the board, and it was difficult to find volunteers to replace them. The professionalisation of NGOs took its course. Holenstein and a group of employees were left to shape the narrative for the coming 1980s.

Holenstein, who had a PhD in literature, never really warmed up to economics. She declared that she did not have the time to study all those models and theories and developed a profound animosity towards (male) economists (Holenstein, Renschler, and Strahm 2008, 35). With Strahm gone, the 'housewives' took over, as Susan George would call them (George 1976, 294–95). The professionalisation of the NGO scene and establishment of labels was dominated by women to a striking extent. It was this group of activist housewives who laid the para-academic groundwork for fair-trade and organic farming labels – the commercial prerequisites for the sustainability narrative (Pestalozzi 1980, 186). They promoted a shift in the narrative towards smallholder agriculture: a simple, eye-catching visual story for the consumer in the supermarket (George 1976, 214).

The smooth transition from grassroots activists to paid NGO employees, from an academic economic concept of fair trade to a narrative on smallholders and sustainability, was first made apparent in George's (1976) "How the Other Half Dies". George, Frances Moore Lappé, author of "Food First" and founder of the Food First think tank, and Holenstein all attended the 1974 Food and Agricultural Organisation of the United Nation's conference in Rome (George 1976, 11; Lappé, Fowler, and Collins 1977). This was the first international conference with a twin-conference set up by NGOs, a concept that would become a custom and effectively stall WTO negotiations after Seattle 1999. Following the para-conference in Rome in 1974, the women activists formed a network of brand-new NGOs, think tanks, and para-academic publications. In her bestseller, George brandished agribusiness, promoted self-sufficient smallholders, and rebranded the Swiss Tanzanian instant coffee campaign. She stopped speaking of Tanzania as a proud African nation that managed to industrialise, and focused on Western activists campaigning for 'villagers' fighting agribusiness:

> "In Switzerland, one group set out to sell Tanzanian coffee, processed in Tanzania and of direct benefit to Tanzanian Ujamaa villagers; they also chose this commodity in order to publicize the unfair trade relationships between first and Third World and the role of Nestles in the control of the coffee trade (George 1976, 294–95)."

After George set the tone by tagging 'unfair trade' in a new way, with smallholders in a village versus a controlling multinational company, Moore Lappé and Holenstein followed up with more books. Holenstein teamed up with Jonathan Power, who left the scene afterwards, similar to Strahm (Power and Holenstein 1976). By the end of the 1970s, Holenstein expanded her already-existing fair trade network with movements for organic agriculture and consumer rights (Holenstein, Renschler, and Strahm 2008, 191–92). On the international level, she took over as coordinator of the food policy study at Johan Galtung's International Peace Research Association in Oslo (Holenstein 1979). This put the small Swiss NGO Berne Declaration at the intersection of a new international movement, which in the 21st century would adopt and promote the term sustainability and focus solely on smallholder farming. To promote this new focus on organic smallholder agriculture, Holenstein collaborated with scientist Joan S. Davis to publish a book for the Swiss public (Holenstein and Davis 1977). In this publication, which was financed by the

World Wildlife Fund and marketed by the Berne Declaration, the two women presented a protectionist program for Swiss agriculture, which has since dominated the narrative:

> *"Demands on agricultural policy*
>
> Legislative measures to protect farms (e.g., direct subsidies, interdiction of industrial livestock farming, tariffs on feed imports) must therefore be supported by consumers in their own interest (Holenstein and Davis 1977, 53)."

Clearly, the two women asked for subsidies for Swiss farmers and tariffs on feed imports. They asked for protectionism in the name of sustainability, but it was protectionism bordering on an idea of autarky, nonetheless. In a host of campaigns typical to the junction between NGO work and the Swiss direct-democratic process, the NGO Berne declaration took part in the construction of Swiss agricultural law. In 1975, they collected 2500 pledges from people in Switzerland declaring they would abstain from eating meat (Erklärung von Bern 1975). They asked for a ban on feed imports, which was partly incorporated in agricultural law.[1] They campaigned for consumers to replace bananas with Swiss apples and published cookbooks with recipes using fair food (Erklärung von Bern 1983). By 1981, the Swiss label for organic food was launched, and Holenstein envisaged an equivalent fair-trade label.

In 1980, the Berne Declaration NGO commissioned a lawyer to conduct a study on a fair-trade label (Adrian Stahel 1980). Originally, the label was to designate unfair products; however, the lawyer discouraged this idea. The next study was commissioned in 1983 by a Protestant charity, and requested that the lawyer lay the groundwork for a neutral fair-trade label (Christoph Lanz 1983). They provided the lawyer with the printed material on fair trade that had accumulated since the Berne Declaration was drafted in 1968. After a thorough review of two decades of printed material on fair trade, in 1983, the lawyer commissioned to write a report on the feasibility of a fair-trade label discouraged food commodity imports. However, considering that the report was based on the original material on fair trade from the 1960s and 1970s, this conclusion was logical. The report stated:

[1] Eidgenössische Volksinitiative «gegen übermässige Futtermittelimporte und Tierfabriken sowie für bestmögliche Nutzung des einheimischen Bodens» (https://www.admin.ch/ch/d/pore/vi/vis130.html) (20.6.2023).

> "Food imports are controversial in terms of development policy. [...] Coffee, tea and rice [...] also prove to be less suitable for a label. [...] Although seemingly obvious, the introduction of a label for food is to be considered rather unfavourable. A label seems more suitable for non-food items (Christoph Lanz 1983)."

However, results of this report were not taken into consideration by the Swiss NGOs of the 1980s, but rather used in a way contrary to the report's intentions. The products the report discouraged (i.e., tropical food commodities) were the products finally chosen to receive fair-trade labels. This development was not unique to Switzerland, but instead a general development that went along with the early volunteer activists leaving the field to be replaced by more donor- and market-oriented employees. While Strahm completely left the development sector and never publicly commented on the strange reversal of his early endeavours, Roy Scott, one of his counterparts in the United Kingdom, the founder of the Oxfam fair trade program Bridge, spoke out in 2015:

> "Bizarre isn't it, you get countries away from cash crops by arguing value should stay with the producer – and now we have the wonderful rigmarole about how great all these food products are – all that is going on is we in Europe import the raw materials and do all the processing (Anderson 2015, 227)."

This 'rigmarole', as former activist Scott called the fair-trade label on tropical food commodities produced by smallholders, was launched in the United Kingdom and Switzerland in 1992 (Anderson 2015). In Switzerland, it became the twin of the organic food label of 1981 in the eyes of consumers. The original academic economic background of the Tanzanian instant coffee sale was long forgotten, and the fair-trade label stood for a narrow range of tropical food commodities produced by smallholders. This was in line with support for local organic smallholder production in Switzerland, which directly led to national agricultural protectionism.

By the mid-1980s, the Swiss flag had crept into the printed material handed out by the Berne Declaration (Figure 3). In 1984, the Berne Declaration urged their donors to support the federal initiative to support smallholder farms.[2] In 1986, the NGO supported the highly protectionist Swiss law on subsidies for beet sugar processing, which would have been unthinkable in 1968 (Moser 1994, 310). The narrative on fair trade had been reversed.

Figure 3. The Swiss flag on a brochure for a federal initiative that the NGO Berne Declaration supported (Erklärung von Bern 1984).

Conclusion

This research shows that Northern development NGOs can be regarded as SIGs in agricultural protectionism. Rodrik's introduction of the 'idea' that trumps interests explains how NGOs complement farm lobbies in the North. Before turning into professional NGOs and SIGs, these organisations were established by activists who promoted industrialisation of the South and market access to the North. It was only during professionalisation that NGOs performed an about-face and started to promote labels for a narrow range of tropical agricultural commodities produced by smallholders.

While Olson's theory of collective action has rarely been applied to development NGOs, our brief historic analysis has shown that it is increasingly useful and valuable to do so. The promotion of family farms in the North and poor peasants in the South is a promising way for the maximization of donations but builds political momentum for a system of

[2] Eidgenössische Volksinitiative «für ein naturnahes Bauern – gegen Tierfabriken (Kleinbauern-Initiative)» (https://www.bk.admin.ch/ch/d/pore/vi/vis167.html). (21.6.2023)

border protections that, paradoxically, prevents peasants to benefit from trade gains.

We conclude it is highly problematic to refer to private, donor-dependent development NGOs' narratives for legislation. When defining PPMs, laws should follow objective criteria, rather than NGO narratives. In economic research, the same scrutiny applied to the role of farm lobbies in international trade should be applied to Northern development NGOs, such as Oxfam or Public Eye. The 'idea' they advocate multiplies the interests of the farm lobby.

Acknowledgement

We thank Jimena Solar Alvarez and Urs Baumgartner for feedback. Funding: This work was supported by the Swiss National Science Foundation [grant number 407340_185603].

References

Adrian Stahel. 1980. Textilien, Mappe 3: Kurzgutachten [Zur Einführung Eines Güte-Schlechte- Oder Labelzeichens in Der Schweiz]. Swiss Social Archives.

AG3W. 1973. Kaffee Ujamaa 73–77: Etikette [Stick-on Label]. Swiss Social Archives.

Anderson, Matthew. 2015. *A History of Fair Trade in Contemporary Britain: From Civil Society Campaigns to Corporate Compliance.* Basingstoke: Palgrave Macmillan.

Anderson, Matthew. 2019. "Charity, Activism and Social Justice: Revisiting Christian Aid's Role in Public Campaigns for Fair Trade, 1968–1973." *Contemporary European History* 28 (4): 535–49.

Banerjee, Abhijit Vinayak, and Esther Duflo. 2011. *Poor Economics: A Radical Rethinking of the Way to Fight Global Poverty.* New York: PublicAffairs.

Baumgartner, Urs, and Elisabeth Bürgi Bonanomi. 2021. "Drawing the Line Between Sustainable and Unsustainable Fish: Product Differentiation That Supports Sustainable Development Through Trade Measures." *Environmental Sciences Europe* 33 (1): 1–13.

Black, Maggie. 1992. *A Cause for Our Times: Oxfam the First 50 Years/ Maggie Black.* Oxford: Oxfam and Oxford University Press.

Brown, Michael Barratt. 1963. *After Imperialism.* London: Heinemann.

Bruycker, Iskander de, Joost Berkhout, and Marcel Hanegraaff. 2019. "The Paradox of Collective Action: Linking Interest Aggregation and Interest Articulation in EU Legislative Lobbying." *Governance* 32 (2): 295–312.

Buchanan, Tom. 2002. "The Truth Will Set You Free': The Making of Amnesty International." *Journal of Contemporary History* 37 (4): 575–97.

Bürgi Bonanomi, Elisabeth. 2015. *Sustainable development in international law making and trade: International food governance and trade in agriculture.* Cheltenham: Edward Elgar Publishing.

Christoph Lanz. 1983. Mappe 3: Ein Entwicklungspolitisches Label – Chancen Und Probleme. Swiss Social Archives.

Collier, Paul. 2007. *The Bottom Billion: Why the Poorest Countries Are Failing and What Can Be Done About It.* New York: Oxford University Press.

Dols, Chris, and Benjamin Ziemann. 2015. "Progressive Participation and Transnational Activism in the Catholic Church After Vatican II: The Dutch and West German Examples." *Journal of Contemporary History* 50 (3): 465–85.

Dümmler, Patrick, and Jennifer Anthamatten. 2020. "Privilegienregister Der Schweizer Landwirtschaft, Aktualisierung 2020: Weiterhin Wachsende Kosten Der Landwirtschaft.".

Eckenstein, Christoph, and Bruno Gurtner. 1977. *Den Dialog Erkämpfen: Industrieländer Und Dritte Welt.* Geneva: Institut universitaire d'études du développement.

Erklärung von Bern. 1968. Einige Zahlen Zur Orientierung. Swiss Social Archives.

Erklärung von Bern. 1969. Bulletin Der Evangelischen Studentengemeinde Zürich. Swiss Social Archives.

Erklärung von Bern. 1970a. International Working Congress of Action Groups on International Development, Egmond Aan Zee. Swiss Social Archives.

Erklärung von Bern. 1970b. Die Schweiz Und Die Entwicklungsländer [Switzerland and the Developing Countries]. Swiss Social Archives.

Erklärung von Bern. 1974. Présentation Du Projet Action Café Ujamaa. Swiss Social Archives.

Erklärung von Bern. [ca. 1975]. Kaffee Ujamaa 73–77, Plakat [Poster on Tanzanian Instant Coffee]. Swiss Social Archives.

Erklärung von Bern. 1975. Protokoll. Swiss Social Archives.

Erklärung von Bern. 1983. GV 1983, Tätigkeitsbericht 1981/82. Swiss Social Archives.

Erklärung von Bern. 1984. Informationsbroschüre Der Schweizerischen Vereinigung Zum Schutz Der Kleinen Und Mittleren Bauern Für Die Kleinbauern-Initiative. Swiss Social Archives.

Franc, Andrea. 2018. "Agricultural Protectionism on the Neoliberal Agenda? The Approach of the Director of the Swiss Business Federation to Agriculture." *Rural History* 29 (1): 81.

Frey, Bruno S. 1985. *Internationale politische Ökonomie.* Wiso-Kurzlehrbücher. Reihe Volkswirtschaft. München: Franz Vahlen.

George, Susan. 1976. *How the Other Half Dies: The Real Reasons for World Hunger.* Harmondsworth: Penguin Books.

Gerlach, Christian. 2015. "Famine Responses in the World Food Crisis 1972–5 and the World Food Conference of 1974." *European Review of History* 22 (6): 929–39. https://doi.org/10.1080/13507486.2015.1048191.

Gessler, Peter. 2008. Persönliche Notizen Zu Den Anfängen Der Erklärung Von Bern Und Ihrem Umfeld. Swiss Social Archives.

Grossman, Gene M., and Elhanan Helpman. 1994. “Protection for Sale.” 84/4. *American Economic Review*, 833–50. http://www.aeaweb.org/aer/.

Haslemere group. 1968. *Haslemere Declaration:* Haslemere group.

Haslemere group. [1969]. *Exploitation of the Third World. Notes on Research and a Bibliography.* London: Haslemere Declaration Group.

Heyes, Anthony, and Steve Martin. 2015. “NGO Mission Design.” *Journal of Economic Behavior & Organization* 119: 197–210.

Hilton, Matthew. 2013. *The Politics of Expertise: How NGOs Shaped Modern Britain.* Oxford: Oxford University Press.

Holenstein, Anne-Marie. 1979. Handakten Anne-Marie Holenstein: IPRA Food Policy Study Group Und World Conference on Agrarian Reform and Rural Development (WCARRD),. Swiss Social Archives.

Holenstein, Anne-Marie, and Joan Davis. 1977. *Zerstörung Durch Überfluss: Überentwicklung - Unterentwicklung Am Beispiel Unserer Ernährung.* Basel: Z-Verlag.

Holenstein, Anne-Marie, Regula Renschler, and Rudolf H. Strahm. 2008. *Entwicklung Heisst Befreiung: Erinnerungen an Die Pionierzeit Der Erklärung Von Bern (1968-1985).* Zürich: Chronos.

Johnson, Erica, and Aseem Prakash. 2007. “NGO Research Program: A Collective Action Perspective.” *Policy Sciences* 40: 221–40.

Jonathan Power. 1969. The Commonwealth and the Common Market. Archives, School of Oriental and African Studies.

Jordan, Grant, and William A. Maloney. 1997. *The Protest Business? Mobilising Campaign Groups:* Manchester University Press.

Krugman, Paul R., Maurice Obstfeld, and Marc J. Melitz. 2015. *International Economics: Theory and Policy.* 10th ed., global ed. Always learning. Boston, MA: Pearson.

Kuhnert, Matthias. 2018. “NGOs, Celebrity Humanitarianism, and the Media.” In *Humanitarianism and Media: 1900 to the Present*, edited by Johannes Paulmann, 263–80 9: Berghahn Books.

Lappé, Frances Moore, Cary Fowler, and Joseph Collins. 1977. *Food First: Beyond the Myth of Scarcity.* Houghton: Mifflin Company.

Malang, Thomas, and Katharina Holzinger. 2020. “The Political Economy of Differentiated Integration: The Case of Common Agricultural Policy.” *The Review of International Organizations* 15 (3): 741–66.

Mazidi, Simon, et al.. “Art. 104a BV – Ernährungssicherheit: Tragweite Und Bedeutung Von Art. 104a Lit. D.” Working Paper.

McAdam, Mark. 2021. *Making Ideas Actionable in Institutionalism: The Case of Trade Liberalization in Kennedy’s Foreign Economic Policy.* submitted.

Möckel, Benjamin. 2019. “Consuming Anti-Consumerism: The German Fairtrade Movement and the Ambivalent Legacy of ‘1968’.” *Contemporary European History* 28 (4): 550–65. https://doi.org/10.1017/S0960777319000262.

Moser, Peter. 1994. *Der Stand Der Bauern: Bäuerliche Politik, Wirtschaft Und Kultur Gestern Und Heute.* Frauenfeld: Huber.

Neidhart, Leonhard. 2002-2014. "Stimm- Und Wahlbeteiligung." In *Historisches Lexikon Der Schweiz*, edited by Marco Jorio. Basel: Schwabe.

Nicholls, Alex, and Charlotte Opal. 2005. *Fair Trade: Market-Driven Ethical Consumption.* London: Sage.

Olson, Mancur. 1965. *The logic of collective action: Public goods and the theory of groups.* Harvard economic studies Vol. 124. Cambridge (Mass.): Harvard University Press.

Pestalozzi, Hans A. 1980. *M-Frühling: Vom Migrosaurier zum menschlichen Mass.* Edited by Migros-Genossenschafts-Bund. Bern [etc.]: Zytglogge Verl.

Power, Jonathan, and Anne-Marie Holenstein. 1976. *World of Hunger: A Strategy for Survival.* London: Temple Smith.

Rodrik, Dani. 2014. "When Ideas Trump Interests: Preferences, Worldviews, and Policy Innovations." *Journal of Economic Perspectives* 28 (1): 189–208. http://www.aeaweb.org/jep/.

Sasson, Tehila. 2016. "Milking the Third World? Humanitarianism, Capitalism, and the Moral Economy of the Nestlé BoycottMilking the Third World?" *The American historical review* 121 (4): 1196–1224.

Schumacher, Ernst Friedrich. 1973. *Small Is Beautiful: A Study of Economics as If People Mattered.* London: Blond & Briggs.

Strahm, Rudolf H. *Effektiver Zollschutz Der Schweiz Und Die Entwicklungländer: Untersuchungen Über Den Effektiven Zollschutz Der Schweiz Auf Produkten Der Entwicklungsländer Und Über Die Auswirkungen Des Schweizerischen Gewichtszollsystems.*

Strahm, Rudolf H., and Christoph Hug. 1975. Z.B. Kaffee Ujamaa. Dossier Zur Verkäuferschulung [Material for Selling Workshops]. Swiss Social Archives.

Thies, Cameron G. 2015. "The Declining Exceptionalism of Agriculture: Identifying the Domestic Politics and Foreign Policy of Agricultural Trade Protectionism." *Review of International Political Economy* 22 (2): 339–59.

Traber, Michael, and Hans K. Schmocker. 1971. *Schweiz - Dritte Welt: Berichte und Dokumente der interkonfessionellen Konferenz in Bern.* Zürich [etc.]: TVZ-Verl. [etc.].

van Dam, Peter. 2016. "Moralizing Postcolonial Consumer Society: Fair Trade in the Netherlands, 1964–1997." *International Review of Social History* 61 (2): 223–50.

van Dam, Peter. 2019. "Goodbye to Grand Politics: The Cane Sugar Campaign and the Limits of Transnational Activism, 1968–1974." *Contemporary European History* 28 (4): 518–34.

Willy, Daniel Kyalo, and Lucy Wangare Ngare. 2021. "Analysis of Participation in Collective Action Initiatives for Addressing Unilateral Agri-Environmental Externalities." *Environmental Science & Policy* 117: 1–7.

Chapter 4

Accelerating the Implementation of the African Continental Free Trade Area Agreement (AfCFTA): Rethinking Youth and the Agriculture Question in Malawi

Elizabeth H. L. Shawa-Mangani,* PhD

Postdoc Research Fellow at the University of Pretoria, Faculty of Humanities, Development Studies, Pretoria, Republic of South Africa

Abstract

Youth unemployment, food insecurity, and poverty are some of the major problems facing Africa. In the region, where there are many millions of young people, agriculture and the rural economy could play a significant role in daily life. There is no denying that young people are essential to the sector and to ensuring that there is adequate food security for the continent. Yet, literature shows that their limited involvement in agricultural activities including rain-fed smallholder farming poses a risk to livelihoods, food production and food security ultimately threatening economic growth and employment creation. This has resulted in a labour deficit in smallholder farming, loss of the most productive and potentially innovative segment of the agricultural labour force. However, timely efforts by the African Union (AU) to accelerate the implementation of the African Continental Free Trade Area (AfCFTA) is a unique opportunity for the continent to refocus its energy and resources on youth empowerment interventions and particularly in agribusiness and trade. Accordingly, this qualitative research which uses Malawi as a case study, contributes firstly, to the contemporary debates on the youth employment

* Corresponding Author's E-mails: u17111472@tuks.co.za; shawaelizabeth@yahoo.com.

In: Agricultural Policy
Editor: Youssef Lee
ISBN: 979-8-89113-720-2

challenge, particularly those centered on the potential role of agriculture in solving youth unemployment, accelerating economic growth and accelerating intra-continental trade. The conclusions emerging from this chapter is a contestation over 'youth are not involved in agriculture' narrative and argues that if barriers are addressed young people can play a significant role in the sector. They can propel the continent's social, economic, and political progress which can transform Africa. There is a need to re-imagine the role of the youth in agriculture and food systems transformation, and how this can significantly impact the reduction of unemployment, poverty, and food insecurity and ultimately accelerate the implementation of AfCFTA.

Keywords: youth, agriculture, food security, transformation, unemployment

Introduction

To say that global youth[1] unemployment is one of the most critical and complex economic and social issues in this time would not be an overstatement. Every year around the world, 40 million young people (400 million in a decade) join a labour market that is not growing enough. Around 70 million out of the 200 million people out of work are young people, and it is warned that, if the global economy does not prove capable of finding a solution, *'we are going to find ourselves with a lost generation'* bringing with it a *'loss of human capital, social exclusion and dislocation'* (ILO, 2019 cited in Shawa, 2020:1).

Africa has the world's youngest population with 400 million people between the ages of 15 and 35, accounting for over 37% of the overall labour force (AU, 2023). Moreover, 65% of the continent's population is under the age of 25, making it the youngest and fastest-growing continent, with a median age of 19.8 years. It is estimated that Africa would be composed of 362 million young people aged 15 to 24 by 2050 (Onah & Okwuosa, 2016; Shawa, 2020). Youth unemployment is a major concern. Joblessness is one of the most pressing development issues confronting governments in the continent, making it perhaps one of Africa's most complex, critical and a big policy concern (Thebe and Shawa, 2022; Xolani, Mlambo and Mkhize, 2022). August (2020) writes that, the aforementioned challenges could be reduced

[1] Youth, youths, young people and young persons will be used interchangeably in this article.

through improved agricultural participation by the youth particularly in the rural areas where farming is practiced (August, 2020: v).

Looking at determinants of youth employment in Africa, empirical studies[2] show that agriculture plays a significantly positive role in generating jobs for young people (Lungu, 2019). Evidence shows that the agricultural sector is a leading employment creator in developing markets and plays a key role in the economic growth and poverty alleviation drive of almost all countries (August, 2020). The sector is not only vital for the development of nations, but also offers support to 60-80% of livelihoods, worldwide, and provides a major boost to national revenue and economic development. The agricultural sector contributes more than 35% in employment creation, worldwide, and profoundly 86.8% in the African continent (ILO, 2017).

Njeru (2017) just like many other scholars (Thebe and Shawa-Mangani, 2022; Fan and Rue, 2020 Chinsinga and Chasukwa, 2012) have observed that although youth have the potential to contribute to the future economy, their limited involvement in agricultural activities including rain-fed smallholder farming poses a risk to livelihoods, food production and food security ultimately threatening economic growth and employment creation. Most young people are leaving and losing interest in farming in developing countries such as Malawi and Zimbabwe (Thebe and Shawa-Mangani, 2022). This has resulted in a labour deficit in smallholder farming, loss of the most productive and potentially innovative segment of the agricultural labour force, affecting agricultural production and food security now and in the future of the continent (August, 2020). Nonetheless, timely efforts by the African Union (AU) to accelerate the implementation of the African Continental Free Trade Area (AfCFTA) comes as a distinctive prospect for Africa to refocus its energy and resources on youth job creation interventions in the agricultural sector through intracontinental trade. Youth can propel the continent's social, economic, and political progress which can transform Africa (AU, 2023).

With the use of a case study, this chapter contributes to debates particularly those centered on the potential role of agriculture in solving youth unemployment, accelerating economic growth and promoting trade. It is a contestation over 'youth are not involved in agriculture' narrative and argues that if youth specific barriers and challenges are addressed in countries like

[2] Thebe and Shawa-Mangani, 2022; Fan and Rue, 2020; Cheteni, P. 2016; Chinsinga and Chasukwa, 2012.

Malawi, young people can play a significant role in agriculture and intra continental trade in Africa.

The African Continental Free Trade Area (AfCFTA)

Since independence, African countries have embraced regional integration as an important component of their development strategies and concluded a number of regional integration arrangements to that effect (Geda and Yimer, 2022 cited on Afrexim, 2022). Among such commitments, the African Union Summit in Kigali, Rwanda, in March 2018 witnessed the signing of the AfCFTA by the heads of state of 44 African nations including Malawi. So far, 36 of Africa's 54 nations have signed the treaty. The AfCFTA aims to establish a single market for goods and services, making it simpler for individuals to move across the continent and foster trade development (Xolani, Mlambo and Mkhize, 2022). The World Bank estimates that the AfCFTA will increase Africa's income by $450 billion by 2035 and increase intra-African exports by more than 81%. According to the UN Economic Commission for Africa, this single market trade agreement will enable the African economy to reach the $29 trillion mark by 2050. As such, Africa has a unique opportunity to lift millions of people out of poverty particularly by empowering youth to change the continent's business environment (ibid). While intra-African trade can serve as a catalyst to promote equity by improving the financial inclusion of youth on the continent, their respective integration into trade policies remains suboptimal (Geda and Yimer, 2022; Ngom, 2023; Xolani, Mlambo and Mkhize, 2022).

Making AfCFTA work for the youth presents an opportunity to address the youth unemployment crisis in the continent and for countries like Malawi. AfCFTA comes as a unique and significant opportunity for the continent to refocus its energy on critical youth led initiatives and particularly in the agricultural sector. The agreement, which entered into force in May 2019, promises to unlock the full potential of Africa's agricultural sector and accelerate sustainable economic growth and food security throughout the region by increasing intra-African agricultural trade which has increased significantly since the early 2000s (ibid). Ngom (2023) notes with optimism that, there is no doubt that this landmark continental trade agreement, if driven by youth, among the most valuable assets of the continent, will be the primary enabler for unlocking Africa's immense potential. However, African governments, policymakers and development organizations must take steps to ensure the agreement reaches its full potential. Herein how the study was conducted.

The Methodology

This section of the chapter outlines the research design, information on criteria used to identify the research data, data collection procedures and tools used to transform collected data into information (data analysis) and how that data was verified to ensure credibility.

Triangulation, Babbie and Mouton (2001), describe a research design as a plan or blueprint of how one intends to conduct a particular study. This is a qualitative exploratory research exercise which used Triangulation. Qualitative methodologies were adopted for they assist in the best way to frame the problems that are under investigation, and make logical links between the problem, data generated analysis and drawing of final conclusions. Qualitative methodologies are preferable for a research such as this as it is based on social inquiry that relies primarily on non-numerical data in form of words, textual and narrative analysis (De Vos, 2012). Triangulation method facilitated validation of data through cross verification from more sources or multiple perspectives. In this case, a case study, document analysis[3] and literature review were used. The documents that were used for this research's evaluation were selected from a wide selection that includes journals, newspapers, memoranda, background papers, brochures, policy documents and books on youth, agriculture, AfCTFA and trade in Malawi and Africa more broadly.

The case study approach was adopted as it allowed for the use of various data sources to explore the youth and agriculture phenomenon within its context of Malawi. In order to obtain detailed, contextualized, and tangible knowledge about a particular practical topic, a case study was a suitable research design. It also provided the study with a 'unit' of analysis given that the issue researched in this study affects a wide range of people. This study uses Malawi as a case to explore the challenges and role of youth participation in agriculture and promoting trade and economic growth. However, it must be acknowledged that, one of the major disadvantages or limitation is that the

[3] Including research data obtained from the author's research PhD project.

findings from a 'unit' of analysis are not necessarily universal (Babbie and Mouton (2001). Therefore, the findings may not be generalized in other contexts and setups.

Thematic data analysis and verification, as stated, the study used Triangulation in order to verify the quality of the raw data and allow for rigorous analysis. Through a desktop analysis and the use of secondary data, the documents used in this research were identified through a systematic search using Google Scholar and a selection of recent academic publications identified through searches of Taylor and Francis Journals, Cambridge Journals and SpringerLink. Thematic data analysis was used in this research. Activities like reading, note-taking, diagramming, and brainstorming to identify themes and sub-themes of the data were involved. Documents were dissected and the contents coded into related topics which were grouped together using a colour scheme. The labels were clustered into themes. The developed themes were analysed and compared to those obtained from related publications. In the end, selected themes were used for the chapter write-up.

Conceptualising Youth, Pathways and Transitions

The years between 10 and 24 is a critical period in a person's life, as rapid changes occur in a person more especially in the earlier years from 10-17. Whilst the exact age at which these changes occur varies according to an individual boy or girl, all these changes happen within a very short period of time[4]. One of the most significant changes is how the person is expected to move into the labour market and become financially independent. Employment is one of the symbols of young people transitioning into adulthood, though the type of employment is critical in determining this. In Malawi, formal employment, with job security and a regular salary, is only one form of employment. Other forms of employment for the youth include seasonal work (ganyu[5]), self-employment with limited job security, and employment in family concerns such as subsistence agriculture.

Subsequently, the world of work is a crucial phase in the lives of young people, with long-term implications for both their individual well-being,

[4] E.H.L. Shawa, 'Youth Unemployment in Post-Democratic Malawi: A Policy Review', (PhD thesis, University of Pretoria, 2020).

[5] Seasonal work and usually not permanent.

society and that of the nation as a whole. Success of other transitions depends on this transition from the world of education to the world of work[6] Young people are arguably facing more 'complex and contested' transitions to adulthood than ever before. For instance, a World Bank 2016 report appealing to lay *'the foundation for mainstreaming youth employment'* describes under employment as a failed or *'unsuccessful'* transition. Accordingly, in many developing countries like Malawi, for both sexes of youth, passage to adulthood is increasingly prolonged, iterative, delayed and partial[7]. Losch (2018:3) demonstrates that, there is a discernible difference with childhood and school age on one side but, on the other side and depending on the social and cultural context, it is possible to be old and to remain in the youth category. It is particularly true in rural Malawi where access to land and to full economic independence can occur late in life.

Losch (2018:4) is of the view that, because agriculture remains central to rural livelihoods, its role in solving the African youth unemployment equation is critical. And its capacity to absorb an increasing labour force is at the same time a big part of the problem and of the solution. In that perspective, the growing disinterest of youth for agriculture appears as a paradox and is probably one of the major issues to deal with. For example, estimations from 2018 reveal that, in South Africa the number of youth in agriculture was not looking good and it decreased by 8% from previous years (Stats-SA, 2020). This positioning faces however a renewed debate among academics, international agencies and governments about the sector offering the best prospects for an inclusive growth process and employment (ibid).

Even so, the International Labour Organisation (ILO)[8] narrates that, there is a tension between the rhetoric which emphasises responsibility for individually and self-consciously navigating youth pathways to adulthood and high expectations of choice, and the reality of available opportunities and imposed constraints, factors which have an impact, but are beyond the control of any individual. This tension may lead to increased frustration and the marginalisation of youths, as well as deeper divisions between those whose opportunities align with dominant discourses and those whose experiences do not measure up to them. Hence, the way young people view their futures needs

---

[6] The World Bank, 'Primary Education in Malawi: Expenditures, Service Delivery and Outcomes' (WDC, World Bank, 2016).

[7] World Bank 2016 report on 'the foundation for mainstreaming youth employment'.

[8] International Labour Organisation, 'Youth transitions and lifetime trajectory' (February, 2019), Available at https://www.ilo.org/ (Accessed on: 4 December, 2023).

to be seen through the lens of the dual epistemology of agency as well as structure. Structure seen as the social landscape or patterned arrangements that influence or limit the choices and opportunities available to any change process, while agency to be seen as the capacity of individuals to act autonomously and make their own free choices, regardless of structural constraints (Moon, 2017). Findings from this study as will be demonstrated further in the chapter shows that in Malawi, youths good decisions and life choices are often not enough, particularly in contexts hampered by history, structural and growing inequalities and policy failures. Overall the findings show an interplay between 'free will' (agency) and 'social forces' (structure). The approach therefore endeavors to deal with what is essentially an age-old problem confronting the social sciences. An attempt to reconcile both structures and human agency as important in the explanation of social life and organization.

Malawi's Economic Growth and the Youth

According to the World Development Indicators Report (2015), Malawi is one of the growing African countries that have at least achieved sustained economic growth over the past decade. It is estimated that the economy grew at an average rate of 6.5% per year between 2003 and 2014 and GDP increased from about USD 7.3 billion in 2003 to USD 13.1 billion in 2014. Nevertheless, the Department for Economic Planning and Development (2017) reported that the country is still categorized as a low-income nation with an estimated GDP per capita of USD 397 in 2014. Persistent poverty, unemployment and income inequality are three of the many development challenges facing Malawi (Shawa, 2020:54). These variables are obviously structural in nature and call for significant changes in the country's growth and development trajectory. This is demonstrated by the fact that despite the country's steady growth numbers over the past decades, Malawi has not seen concomitant improvements in the material conditions of existence of the vast majority of her citizens and especially the youth. Underemployment (about 27%) is a major problem in Malawi's labour market. This refers to casual, non-frequent job opportunities that do not meet the most basic needs of those so employed. In addition, the productivity of Malawi's labour force is undermined by morbidity, premature mortality and poor economic skills, which stem from low levels of education and insufficient opportunities for skills development amongst the youth. The argument is that a key reason for not achieving inclusive, pro-poor and job-enhancing growth is the country's narrow economic base (Department for Economic Planning and Development, 2017 cited in Shawa: 2020: 55).

Notwithstanding the above, to date, agriculture remains the backbone of the Malawi economy. Even so, although agriculture is officially described as an engine of economic growth and is the largest employer in Malawi, efforts to comprehensively address the question of persistent youth unemployment have been few and far in between for the past decades. Apart from young people's disinterest (agency), Chinsinga and Chasukwa (2012) the main problem is that the growth of this sector is largely on autopilot and not a consequence of any strategic government inputs in driving such growth (structure). For instance, Chinsinga and Chasukwa in 2018 further highlights perhaps harshly that, there is no information or systematic investment, the road network and infrastructure is poor, and so too, the creation of markets and commodity sales, low commodity prices, unfavourable trading laws and minimal government assistance in the agricultural sector more generally, all these are prevalent challenges which demotivate young people. The social structure of the sector does not attract the youth. The section next will dive deeper into the youth and agriculture question in Malawi.

Malawi Youth Participation in Agriculture: Perceptions and Aspirations

Babbie[9]shows that, at a macro level, agriculture is the 'backbone' of many Sub-Saharan countries where it accounts for 37% of the GDP. To date, agriculture remains the pillar of the Malawi economy (Chinsinga and Chasukwa, 2017). However, there are a number of stereotypes that young people attach to the sector in the country.

The dominant perception among youth in Malawi is that the agricultural sector, sub-sectors (and related industries) create job occupations mostly for the illiterate. Neither do many youth, especially the educated, regard agricultural services as a business area worth investing time and energy in. This could mostly be the case because every year they are saturated with

[9] Babbie, K. (2017) article on 'Rethinking the *'Youth Are Not Interested in Agriculture'* Narrative'.

stories of the dysfunctional agriculture markets that rip off farmers due to government's inability to enforce minimum produce prices[10]. The social landscape or patterned arrangements limit the choices and opportunities available for young people in the agricultural sector in Malawi, as earlier indicated that structure and non-vital life events play a critical role in influencing young people's life choices.

Similar observations were also made by Chinsinga and Chasukwa (2017) who writes that, in Malawi, graduates are reluctant to work in rural areas because of poor facilities and many are hesitant to work in the public sector because of low pay. They are what they call the *"network"*[11] generation which cannot imagine living in areas without electricity, where phones cannot work, where internet is inaccessible and roads are in a bad shape. Chisinga and Chasukwa (2018) emphasised that, Malawi youth conceptions and aspirations mostly revolved around

> Owning a decent iron sheet roofed house, ready availability of food and fashionable clothes for our families, sending children to a good boarding school and at least having a mountain bike and a television screen. The perception is that a person can spend all his or her time in agriculture but still be very poor while those who are educated and bold enough to venture and migrate out to the outside world can prosper within a very short period of time (ibid).

In connection to the above, indeed research shows that youth in Africa tend to demonstrate higher willingness to move permanently to another country: 38% in Sub-Saharan Africa and 35% in Northern Africa in 2015, (globally the figure stands at 20%). For instance, among Sub-Saharan African countries, the percentage of youth willing to migrate ranges from 77% in Sierra Leone to 11% in Madagascar[12]. The International Organisation for Migration (OIM) in 2016 did acknowledge that in the minds of these young people, migration represents an opportunity for youth to provide a better life for themselves and their families, pursue educational aspirations, improve their professional skills prospects, or satisfy a desire for personal development

---

[10] Youth Decide, 'Malawi National Youth Manifesto 2019 – 2024', *Consortium of civil society Youth-NGOs in Malawi* (Lilongwe, Project4 Digital Design, 2019).

[11] Technology driven.

[12] E.H.L. Shawa, 'Youth Unemployment in Post-Democratic Malawi: A Policy Review', (PhD thesis, University of Pretoria, 2020). Page 185.

and achieving their life aspirations[13]. As such their personal aspirations and willingness to dive into the agriculture sector are shaped against this background. Young people do not see agriculture as a means to achieve their life aspirations (Shawa, 2020).

In a similar manner, Malawi's historical context points to migration. Most young people in the country, particularly those from rural settings attach *"the good life[14]"* to migrating across boarders or urban areas. For these youths' migration represents

> If you ever find a decent and nice looking house with a solar panel, plasma television screen and a mountain bike, it is not because of farming, the owners acquired them through other means, such as working in South Africa or Zimbabwe and Mozambique and while there sent money to their wives and relatives to help them build houses with burnt bricks, roofed with corrugated iron sheets with DSTV[15].

It simply adds that, even though Malawi's youth are typically in transition, navigating many identities and responsibilities, taking up farming as a profession has rarely captured the imaginations of upwardly mobile young people whose dreams and search for opportunities takes them to mostly white-collar professions, mainly in cities[16] such as Lilongwe[17] or out of the country to countries such as South Africa.

To add, in an increasingly globalised world with fast-evolving communication and media technology, young people in remote areas of Malawi are ever more aware of urban-rural inequalities and aspire to standards of living not typically associated with agricultural or rural livelihoods. For instance, social media such as Facebook and WhatsApp contribute to aspiring for a life that is considered outside of agriculture. Social reality influences beliefs, attitudes and behaviours of these youths. Stereotypes are constantly used in agriculture. These are normally exacerbated by what books, posters, media says...

---

[13] Ibid.

[14] Life aspirations.

[15] Focus Group Discussion in 2018 cited in Shawa, 2020.

[16] The World Bank, 'Primary Education in Malawi: Expenditures, Service Delivery and Outcomes' (WDC, World Bank, 2016).

[17] The capital city of Malawi.

> Everyone thinks it's the farmer with the overalls, with a wheelbarrow, and brown dirty torn and shirt that grows all the food, and takes care of the animals is what farming and what you will end up to be if you are in agriculture (Focus Group Discussions, 2018 cited in Shawa, 2020).

Nonetheless, scholars have underlined that, perceptions of accessibility depend on the information to which individuals are exposed and the information they search for. McDonald et.al (2014) determined that intentions to participate in an activity could be predicted based on a person's knowledge, observations or other information held about some issues or event. Hence for example in school the decision to undertake agriculture as a course or a career is predicted by examining their beliefs about agriculture. McDonald et. Al (2014) further indicates that in Africa, agriculture employs 65% of Africa's labour force and accounts for 32% of GDP. However, unlike the United States, it is dominated by family farms and the off-farm sector is very weak and there has been little improvements in yields and production techniques. Due to these structural differences, there are bound to be divergence in the literatures and empirical studies on youth's aspirations and perceptions of agriculture in developed and less developed countries such as Malawi. Despite the differences, it can be observed that apart from structural opportunities, family and parents influence youth aspiration and perception towards agriculture. Most of the literature on young people and agriculture acknowledge the important role of parents and family members in influencing young people's aspirations to be involved in agriculture sector or vice versa. In the African literature, parents who have experience working on farms are less keen to let their children do likewise because they are aware of agriculture risks and low returns. Tafere and Woldehanna (2014) find that parents in Ethiopia for example have a strong influence over their children's occupational aspirations, and that both parents and children alike prefer non-farming to farming occupations. That study revealed that at age eight, most children aspire to become teachers when they leave school whereas more than 40% want 'high' and 'other non-farming' occupations, leaving only a small proportion who want to farm (ibid). This study confirms that, another factor that influences young people in Malawi not to take up farming is their parents who hold negative attitudes towards agriculture even if they are successful farmers.

Overall, Chinsinga and Chisukwa (2012) emphasise that the combination of many factors has given agriculture a bad image among the young people as a primary source of livelihood. Most of the young people describe working in the agricultural sector as *'dirty work and demeaning'*. Others consider agriculture to be *'simply the basic means of survival'*. For the young people in the country, therefore, there are three alternatives as is also observed in this study through which they can realise their dreams of *"the good life"*: (1) migrating to urban areas such as Lilongwe, Blantyre and Mzuzu[18] in search of employment; (2) engaging in small scale business; and (3) migrating to South Africa. These are considered the surest means of attaining *'the good life'*, while agriculture is condemned as laborious, less rewarding, exploitative and requiring a long time to reap rewards. The above could symbolise that young people in Malawi do not have the agency to take up agricultural opportunities due to the negative stereotypes that is attached to the sector in Malawi.

However, Babbie (2016) reported that, the youth and agriculture question runs deeper than the 'youth are not interested' narrative. The attitudes and agency of the youth are shaped primarily by the significant barriers they face in accessing the capital needed to develop agriculture-based enterprises, all these conditions must be in youth policy debates.

Agricultural Policy and Strategies in Malawi: Youth Challenges in Agriculture

Chinsinga and Chasukwa (2017), writing in a policy brief titled 'Agriculture and Youth Unemployment in Malawi', states that, in Malawi, the youth should be key in the agriculture sector which is not only the backbone of the country's economy but also an essential part of the social fabric. This is, however, not the case. The youth are not actively involved in the agricultural sector, apart from the harmful and exploitative participation as child labourers and in insecure work opportunities. One of the reasons could be that the policy environment is not favourable for young people (Chinsinga and Chasukwa, 2017). Although agriculture is officially described as an engine of economic growth and is the largest employer in the country, efforts to comprehensively address the question of persistent youth unemployment have been few and far in between. As earlier alluded by Chinsinga the main problem is that the growth of this sector is largely on autopilot and not a consequence of any strategic government inputs in driving such growth. One of the important gaps in most national policies, for instance, is that while the proportion of women

[18] The main urban cities in Malawi.

and youth in waged employment in the non-agricultural sector is used as an indicator to judge the success of increasing opportunities for women, no specific mention is made of young people (ibid: 2017). As such, there is a call to create a conducive policy environment for the youth in Agriculture and the National Youth Policy should be the central point of reference for these policy changes (Shawa: 181).

Most policies including youth policies such as National Youth Policy (2023 – 2028) and programmes in Malawi do not pay explicit attention to the agricultural sector even though its central theme is the promotion of youth empowerment. Agricultural policies and programmes offer the youth minimal opportunities in any agriculture related trade. This would pose as a serious hindrance in the implementation of the AfCFTA. According to Chinsinga and Chasukwa (2017), the problem of youth unemployment in the country can be successfully dealt with when serious attention is paid to forward and backward linkages between youth policies and the agricultural sector, in ways that cover the entire value chain of possibilities in this critically important sector. It is also the one sector that could potentially lay the foundation for high growth, diversification and expansion to other sectors, sub-sectors and industries. This is exactly how the high-growth economies such as South Korea, Taiwan and Malaysia were initiated (Moon, 2017). Further findings are noted:

> The apparent neglect of agriculture as a potential source of employment for the youth could mainly be due to the political imperatives associated with a democratic political dispensation in a neo-patrimonial setting whereby the state functions largely as a transfer pump of resources to the favoured sections of the community. Political elites are often preoccupied with devising strategies that can maintain them in power in the five-year electoral cycles. In this scheme of things, the youth are often prioritized except for being captured as instruments of terror by the politicians (Chinsinga and Chasukwa, 2017:2).

The Fertiliser Subsidy Programme (FISP) for example, with the potential of increasing food security among smallholder farmers excludes the youth as its beneficiaries. The main reason for their exclusion is that the youth do not own land and are energetic enough to work elsewhere to generate income to purchase agricultural inputs on their own from the market. Yet the obsession of politicians with FISP as a vote-spinning machine has made it impossible to shape its implementation dynamics in a way that it would make the agricultural sector in general more productive and attractive to young people (ibid:3). In connection, the government did set up the Youth Development Credit Scheme (YDCS) and the Youth Enterprise Development Fund (YEDEF), specifically to provide financial support to young people interested in setting up businesses. However, as Chinsinga and Chasukwa (2017), have demonstrated, perhaps harshly, both the YDCS and the YEDEF have not been very effective or efficient. Though they were both championed as youth initiatives, the primary targeted youths were basically those affiliated to the parties in power. As a result, both initiatives have failed to pass the sustainability test (ibid).

Chinsinga and Chasukwa (2017), writing further on 'Agriculture and Youth Unemployment in Malawi' acknowledge that the discontinued Youth Job Creation Initiative (YJCI) launched by President Joyce Banda on 15 March 2013 appeared poised to bring back agriculture to the centre of youth development. The YJCI was launched as a strategy for dealing with the problem of chronic youth unemployment by exporting unskilled, skilled and semi-skilled labour to countries where they could acquire skills and knowledge that would in turn be put to productive use upon their return. The target countries for youth labour export were South Korea, Kuwait, and the United Arab Emirates. The programme had huge potential to contribute to the development and subsequent transformation of the agricultural sector. The agricultural sector in Malawi is deemed to be relatively backward and the exposure of a good number of the youth to South Korea's advanced agricultural system could have, perhaps, contributed to the drive to modernizing the country's agricultural sector (ibid). There is a need to rethink the role and place of agriculture, as a driver of job-creation and general youth development in the NYP (Shawa, 2020: 183).

Furthermore, Young people do not have information, for example to access tenders floated by government/companies, not to mention that taxes are generally high, which tends to erode the little that may be generated in agribusiness. Most policies favor international investors, especially large multinational firms rather than young locals. Consequently, space for real youth empowerment, growth and entrepreneurial success are stifled (UNDP, 2015 cited in Shawa 2020).

To add, young people are capable of great collaboration and promote of industrialisation which can lead to increased African agricultural innovation and entrepreneurship hubs. This in return can increase skills and knowledge exchange which is essential for the implementing of the free trade area zone in the continent. Through expanded access to markets, youth have a major role to play in diversifying the sources of growth and exports which can strengthen the implementation and impact of the AfCFTA (Ngom, 2023). In Malawi the National Planning Commission addressing 'Youths in Trade and Industry: Opportunities and Challenges', stressed the power of collaboration and called on the youths to come together and work through partnerships or agricultural cooperatives to explore different opportunities in Malawi. Research shows that majority of enterprises in Malawi are one-man shows, without even any employee. This makes the businesses weak and without capacity to explore opportunities. (Zhuwao, 2022 cited on NPC, 2022).

The findings also show that young people can hold decision-makers accountable for the commitments they have made towards sustainable development. As such being key stakeholders of AfCFTA could be an accountability tool. In Malawi there is a need to emphasize on enhancing governance and accountability in the use of public resources and efficiency in service delivery, reinforcing anti-corruption institutions and systems, and empowering youth (Shawa, 2020). As illustrated above initiatives such as YEDEF and YDCS have failed due to political interference (Chinsinga and Chasukwa, 2018). In the country, it is the reality of democratic politics that they operate and are often driven around five year electoral cycles. While this time honored requirement of democratic systems is good for accountability and citizen participation in the definition of its country's future, it can and has often undermined policy consistency. It is common cause that a new administration or a new minister (even from the same political party) will do their best to introduce 'new' policies and discard the 'old' policies of their predecessors. In some instances, a different political party does not want to be associated with policies of the previous regime. This policy zig-zag has had devastating consequences on Malawi's youth development performance including the agriculture sector (Shawa, 2020: 229).

In line with good governance, Chidede and Bore (2022) argues that trade supported by improvements in trade governance can contribute to reducing inequalities by distributing benefits more equally across all segments of the society. They further suggested that, meaningful participation by all groups of society including marginalized groups such as youth in trade policy making processes can contribute to achieving inclusive, sustainable growth and development, as well as social stability in Africa.

The Role of the Youth in Agriculture and Implementing AFCFTA

Data indicates faster growth in imports from the rest of the world than from intra-African sources. Hence, while there has been an absolute increase, market share has remained flat, suggesting that the continent has yet to capture a growing slice of the rapidly growing demand. This should be the ultimate focus of and measure of success for the AfCFTA. Africa's large and fast-growing youth population is considered one of its greatest assets, with a central role to play in shaping the development of the continent. The significance of the AfCFTA cannot be overstated: It will be the world's largest free trade area since the establishment of the World Trade Organization (WTO) in 1994 (Signé and van der Ven, 2019). Harnessing the power and creativity of Africa's youth to accelerate the transformation of African economies is one of the most critical issues in implementing AfCFTA. Young people should be drivers and beneficiaries of agribusiness and international trade. Why should this be the case?

According to the International Trade Centre (ITC), 'youth are 1.6 times more likely to start a business than those above 35 years old' (WTO, 2019). Findings from this study shows that in Malawi young entrepreneurs are an important prospect for the country's economy. Over 54% of young people are self-employed in the country. If favourable conditions are in place, entrepreneurship can be a pathway to decent work and have tremendous impacts on the economy as it can boost trade. Many new businesses are created in developing countries, which on average has more nascent and young companies than established firms (Kew et al., 2013 cited in WTO, 2019). Young people are most likely to be involved in trade more than adults. They have the zeal and are easily willing to start-up businesses which can contribute to a large market area for AfCFTA.

In the context of Malawi, findings show that strategies must be implemented that aim to promote agribusinesses, access to finance for youth and other vulnerable groups in agriculture, in particular by linking them to financial services and agricultural credit. Specific for Malawi, youth cooperatives must be encouraged to enable youth sell produce at competitive prices as well as promotion of production of high value agricultural produce such as flowers. Vocational education and skills training focused on the possibilities available in agriculture is long overdue and so too, the need for the creation and expansion of value chains that can help to catapult the country from a narrow agri-based economy to a diversified, high-growth economy, with endless value chains and with more youth participation in agriculture sector it will ensure farming is passed from one generation to the next (Shawa, 2020: 163).

Studies show that young people have the power to create innovative home-grown powerful solutions which can create new jobs and strengthen competition in the country. Through the establishment of new businesses and inventive endeavors, young entrepreneurship contributes to the GDP growth rate of nations by bolstering national productivity. Part of the AfCFTA agreement is that young people can receive assistance from the financial sector if they are prepared to take on the role of entrepreneurs in the agricultural sector. In general, youth entrepreneurs can increase a countries' investment through their involvement in agriculture (Shawa, 2020; Thebe and Shawa, 2023; Signé and van der Ven, 2019). Outcomes in the study shows that young people should be given low-interest access to finance which does not require property-based collateral and guarantees. Access to microfinance remains a big challenge for the young people in Malawi. Youth fail to secure loans because most lending institutions insist on collateral, which youth in their predicament of multiple vulnerable factors cannot provide (Shawa, 2020). It has become a double blow; youths cannot access a bank or microfinance loan and cannot access a government development loan[19]. Furthermore, there seems to be a stereotype that the intervention of youth in rain-fed smallholder agriculture in Malawi is associated with high costs, as you have to provide young people with loans, machinery, land and equipment. A cost the government is not willing to succumb to. The inaccessibility to financial capital is one of the major problem hampering the participation of youth in agriculture. With that lack of finance, young people either lose their interest or think of it as a last resort (ibid). For a successful implementation of youth agricultural trade initiatives through intra-continental trade, an enabling legal, regulatory and administrative environment must be created for the youth of the country to actively participate in trade through AfCFTA.

It has now become common knowledge to know that youth are more energetic, daring and innovative compared to adults. Globalization and digital technology has played a huge part in helping youth to be more innovative than adults. Young people are not afraid to brake the status quo, and they are not

[19] In reference to YEDEF and YDCS.

limited to cultural barriers. Young people are often innovators in different industries (Shawa, 2020). As such as AfCFTA is about E-commerce and the digital economy, young people can be critical implementers. As Malawi 2063 policy by the Malawi National Planning Commission (NPC) states that, digital technology is a powerful enabler as it could open up many opportunities since an investment in ICT is also an investment in economic growth, jobs, education, health, agriculture, and good governance (NPC, 2022). This must be reinforced in order for Malawi to achieve the visions of the policy.

Most important, Ngom (2023) is of the view that Africa will not take full advantage of the AfCFTA if it remains the epicenter of humanitarian action. It is said that, today more than ever our continent has been faced with multiple humanitarian crises including climate change, wars, food insecurity, disease outbreaks, poverty, inflation, conflict and worsening health inequity(ibid). Young people's essential role in peace and security has been increasingly recognized and documented since 2015. The implementation of AfCFTA requires good governance, peace, and security to be successful. It is critical to ensure that peace and security are maintained throughout the implementation of AfCFTA as conflict and instability can negatively impact and hinder trade and economic growth. Youth have a role to promote peace and security to accelerate trade and economic growth within the continent (ibid).

In all, young people have a high chance and likelihood to start a business which could contribute to job creation and curbing the rising rates of unemployment. The youth can bring in new innovation as most are tech savvy which can help in e-commerce. Youth are capable of great collaboration and promote agribusinesses and industrialisation which can expand the African agricultural market. Addressing their barriers and acknowledging them as key stakeholders in implementing AfCFTA, young people can propel the continent's social, economic, and political progress which can transform Malawi and Africa at large.

Conclusion and Implications

Young people should be drivers and beneficiaries of agrifood markets, agricultural trade and food security as their potential can be harnessed to achieve the continent's sustainable development. Despite making up a significant percentage of the continent's population, young people's participation in agricultural matters is still very limited in most developing countries including Malawi.

In this chapter, Malawi was used as a case study. Findings demonstrate that youth participation in agriculture has become a focal point at the centre of government's strategic plans. However, young people in developing countries such as Malawi continue to face a wide range of bottlenecks that constrain their capacity to participate and benefit in agriculture and international trade. This has had major implications on the continent and globally, it has given rise to significant setbacks for continental food security and prosperity.

On one hand, some of the specific structural barriers in agricultural trade include lack of access to assets, capital and finance; poor governance and political interference, limited access to trade-related or market information, limited trade-related education or skills; and hostile environment and administrative and regulatory frameworks that do not favor young people. On the other hand, most of the youth in the country do not see agriculture with all its inherent challenges, particularly its low returns and a very long gestation period as a means of attaining their version of *'the good life'*. Due to their perception of farming being *'dirty, antiquated, unprofitable and for the elderly and uneducated'*. The sector is seen as less rewarding and that their aspirations and *"the good life"* cannot be attained. Agency and own free choices playing a role to these aspirations.

Future policies and strategies must endeavour therefore to deal with what is essentially an age-old problem confronting the social sciences. This is whether individual actors (or group of actors), youths of Malawi in this case, act or can act in a 'constraints-free and structure-less policy environment, or whether institutions and structures always influence, shape and determine the possibilities of such change. Only a policy approach that factors in the role of both structure and agency is able to navigate this layered and complex maze, with multiple variables that shape and influence each other.

Policy makers need to implement supportive policies and investment to ensure youth participation. There is a need to align agricultural strategies and interventions with youth career aspirations in the country. It is critical that policymakers and practitioners developing such interventions understand youth imagined futures in relation to perceptions, preferences and the quality of life to which they aspire.

The main conclusion emerging from this research is a contestation over 'youth are not involved in agriculture and trade' narrative and argues that if barriers are addressed young people can play a significant role. Young people have a high chance and likelihood to start a business which could contribute to job creation and curbing the rising rates of unemployment. The youth can bring in new innovation as most are tech savvy which can help in e-commerce and new technology in agriculture. Youth are capable of great collaboration and promote industrialisation which can expand the African market. Addressing their barriers and acknowledging them as key stakeholders in implementing AfCFTA, young people can propel the continent's social, economic, and political progress which can transform Africa and help achieve the 'Africa We Want' as stipulated in Africa's Agenda 2063.

References

Africa Union. (2023). *Youth Development.* Addis Ababa, Ethiopia.

Afrexim Bank. (2022). *Making the AfCFTA Work for the Africa we Want.* Available at https://www.afreximbank.com/making-the-afcfta-work-for-the-africa-we-want/ (Accessed on: 19 October, 2023).

August, A. A. (2020). *Youth's Aspirations and Perceptions towards Agricultural Participation: A Case of two Free State Regions.* Masters Thesis, University of Free State, Bloemfontein. Available at https://scholar.ufs.ac.za/bitstream/handle/11660/11241/AugustMM.pdf?sequence=1&isAllowed=y (Accessed on: 2 February, 2024).

Babbie, K. *'Rethinking the 'Youth Are Not Interested in Agriculture' Narrative',* (2017). Available at https://nextbillion.net/rethinking-the-youth-are-not-interested-in-agriculture-narrative/, (Accessed on: 1 July, 2023).

Cheteni, P. (2016). *Youth Participation in Agriculture in the Nkokobe District Municipality, South Africa.* 55(3): 207-213.

Chidede, T. and Bore, O. (2020). *Youth in trade and trade governance in Africa.* Available at https://www.tralac.org/blog/article/14727-youth-in-trade-and-trade-governance-in-africa.html#:~:text=This%20youthful%20population% |20can%20be,as%20the%20continent's%20biggest%20resource (Accessed on: 15 March 2017).

Chinsinga, B. (2018). The Green Belt Initiative, Politics and Sugar Production in Malawi, *Journal of Southern African Studies,* 43:3, 501-515.

Chinsinga, B. and Chasukwa, M. (2017). *PAS Policy Brief: Agriculture and Youth Unemployment in Malawi.* Department of Political and Administrative Studies (PAS). March 2017 Volume 1 No. 2. Chancellor College, Zomba.

De Vos, A. S. (ed) (2012). *Research at Grass Roots: For the Social Sciences and Human Service Professions.* Pretoria: Van Schaik.

Shawa, E. H. L. (2020). *'Youth Unemployment in Post-Democratic Malawi: A Policy Review',* (PhD thesis, University of Pretoria, 2020).

Fan, S. and Rue, C. (2020). The Role of Smallholder Farms in a Changing World. In: Gomez y Paloma, S., Riesgo, L., Louhichi, K. (eds) *The Role of Smallholder Farms in Food and Nutrition Security.* Springer, Cham. Available at https://doi.org/10.1007/978-3-030-42148-9_2 (Accessed on: 21 June 2022).

Government of Malawi, *'Malawi Youth Status Report 2016: Adolescent and Youth Situation Analysis* (SitAn)' (Lilongwe, Government Press, 2016).

International Labour Organisation, *'Youth transitions and lifetime trajectory'* (2019). Available at https://www.ilo.org/ (Accessed on: 4 June, 2022).

International Labour Organisation. (2019). *'Youth transitions and lifetime trajectory',* Available at https://www.ilo.org/ (Accessed on: 4 June, 2022.

Losch, B., Freguin-Gresh, S. and White, E. (2012). *Structural Transformation and Rural Changes Revisited: Challenges for Late Developing Countries in Globalization World.* Washington, DC: World Bank.

Lungu, I. (2019). *A Fresh Chance for Africa's youth: Labour Market Effects of the African Continental Free Trade Area (AfCFTA).* Available at https:// www.giz.de/de/downloads/Policy%20Paper%20-%20Labour%20Market %20Effects%20of%20the%20AfCFTA.PDF (Accessed on: 22 November, 2023).

McDonald, P. Bailey, J. Pini, B. and Price, R. (2011). 'Young people's aspirations for education, work, family and leisure', *Work, Employment and Society,* Volume 25 (2011), Issue 1.

Moon, S. (2017). *Political Economy of State-making in Post-apartheid South Africa.* New Jersey: Africa World Press.

National Statistical Office. (2014b). *Malawi Labour Force Survey 2013. Zomba.* National Statistical Office.

National Planning Commission (NPC) of Malawi. (2022). *Youths outline opportunities, challenges in trade, industry.* Available at https://npc.mw/ 2022/01/youths-outline-opportunities-challenges-in-trade-industry/ (Accessed on: 01 December, 2023).

Njeru, L. K. (2017). Youth in Agriculture: Perception and Challenges of Enhanced Participation in Kajiodo North Sub County, Kenya. *Greener Journal of Agricultural Sciences 7* (8): 203- 209.

Ngom, M. (2023). *AfCFTA: Reaping the benefits of the world's most youth and women-friendly trade agreement.* Available at https://www.un.org/ africarenewal/magazine/february-2023/afcfta-reaping-benefits-world% E2%80%99s-most-youth-and-women-friendly-trade-agreement (Accessed on: 23 September, 2023).

Onah, N. G., & Okwuosa, L. N. (2016). Youth unemployment and peace in Nigerian society. *Mediterranean journal of social sciences,* 7(1 S1), 52.

Shawa, E. H. L. (2020). *Youth Unemployment in Post-Democratic Malawi: A Policy Review,* PhD Thesis, University of Pretoria, Pretoria. Available at http://hdl.handle.net/2263/74453 (Accessed on: 2 February 2022).

Signé, L. and Van der Ven, C. (2019). *Keys to success for the AfCFTA negotiation. Africa Growth Initiative. Policy Brief.* Available at https://www.brookings.

edu/articles/keys-to-success-for-the-afcfta-negotiations/ (Accessed on: 01 December, 2023).

Statistics South Africa (Stats SA) (2018). *Commercial Agriculture Report.* Pretoria: Stats SA.

Tafere, Y. (2014). *'Education Aspirations and Barriers to Achievement for Young People in Ethiopia',* (UK: Oxford Department of International Development (ODID), 2014).

Thebe, V. and Shawa-Mangani, E. H. L. (2022). Youth, Livelihoods, and Complex Realities in a Former Migrant Labour Reserve in North-western Zimbabwe, *Journal of Applied Youth Studies.*

Thebe V. (2018). Youth, Agriculture and Land Reform in Zimbabwe: Experiences from a Communal Area and Resettlement Scheme in Semi-arid Matabeleland, Zimbabwe. *African Studies* 77(3):336-353.

The World Bank, (2016). *'Primary Education in Malawi: Expenditures, Service Delivery and Outcomes'* WDC, World Bank.

World Trade Organisation. (2019). *Empowering Youth for Sustainable Trade.* Available at https://www.wto.org/english/res_e/booksp_e/aid4trade19_chap8_e.pdf Accessed on: 12 October 2023.

World Bank. (2020). T*he African Continental Free Trade Area: Economic and Distributional Effects.* Washington D.C.: The World Bank.

World Bank. (2009). *African Development Indicators. Youth and Employment in Africa: The Potential, the Problem, the Promise.* World Bank, Washington, D.C.

Xolani, T., Mlambo, V. H., and Mkhize, N. (2022). The African Continental Free Trade Area Agreement (AfCFTA): Possible benefits for women and youth in Africa. *Latin American Journal of Trade Policy* 13 (2022) - ISSN 079-9668 - Universidad de Chile.

Youth Decide, (2019). 'Malawi National Youth Manifesto 2019 – 2024', *Consortium of civil society Youth-NGOs in Malawi*, (Lilongwe, Project4 Digital Design, 2019).

Chapter 5

A Cursory Glance on Agricultural Policy Analysis in a Changing Landscape

Samir Mili*
Institute of Economics, Geography and Demography (IEGD),
Spanish National Research Council (CSIC), Madrid, Spain

Abstract

Policy interventions on bio-natural and socioeconomic structures cannot be explained and predicted satisfactorily through simple mental constructions. Formal analysis is necessary to coherently represent complex relationships linking agricultural activities to their socioeconomic, environmental and political impacts. Formalization provides explanations of key logics and causal mechanisms for the sequential stages of agricultural policy processes, and allows their controlled assessment. In this context, the present paper provides a brief scrutiny of major issues in agricultural policy analysis aiming at contributing to informed policy debate in this subject area. A global framework is drawn up for market structures and policy measures. Prominent agricultural policy issues including environmental sustainability, poverty reduction, developments in global trade, as well as main features of techniques used in analyzing these issues are briefly depicted.

Keywords: policy analysis, policy effects, agriculture, international trade

---

* Corresponding Author's Email: samir.mili@csic.es.

In: Agricultural Policy
Editor: Youssef Lee
ISBN: 979-8-89113-720-2

Introduction

Agriculture has always been a strategic sector in the vast majority of countries. Its critical importance has increased in the last few years with soaring food prices and market volatility, aggravated by the Covid-19 pandemic, climate change challenges and multiple geopolitical crisis, which have endangered the food security of many countries. Developments in agricultural production and markets are highly sensitive to public policies (Bastidas-Orrego et al. 2023). Indeed, it is a long-established evidence that agricultural policy measures have a significant influence on the welfare of agents involved in agricultural production and trade. Such measures are designed and implemented for multiple reasons and motivations. Research attempting to evaluate and predict policy impacts has been in high demand due to increasing policy changes in agriculture. After the almost exclusive focus on trade issues during the 1980s and 1990s, evidence showed that other relevant issues also deserve urgent, sustained attention of analysts. These include sustainable development, employment, migration trends, climate change, disruptions of value chains for both agricultural commodities and key inputs (OECD 2023a), and their ensuing effects on income distribution and household poverty, especially in developing countries where production and trade structures differ significantly from those in developed nations.

Much applied research was already available, however since policy measures affecting the domestic markets of the developed nations were going to spill over to international commodity markets, the issue affected the interests of the whole international community. In consequence, the debate moved progressively to multilateral forums and agencies dealing with trade across borders (WTO, World Bank, regional development banks, UNCTAD) and to a long list of academic organizations. In this process, many researchers and experts have been producing advice and analytical instruments attempting to assess the possible effects of the proposed measures.

Undeniably, political and socioeconomic developments are often inter-related through subtle links of trade connections, policy decisions, structural constraints, and even beliefs and cultural aspects. As a result, policy scenarios are far too complex to be analyzed by simple mental constructions. Formal modeling offers the possibility of considering the relationships amongst a large number of variables and of simulating different outcomes when some of these variables change or interact. In this context, while modeling activities have reflected the enormous flexibility and imagination of model builders, results achieved a varied degree of success when the instruments were applied to the realities of advanced countries following the general premises learned about their economic structure. However, it became obvious that the particular conditions of developing countries need to be taken into account if the usefulness of modeling for understanding economic processes in developing countries was to be retained. Indeed, when modeling policy-driven situations in developing countries, some important features must be kept in mind. First, there are differences in priorities. It might seem obvious, but it is sometimes forgotten that, in many poor countries, poverty is the most pressing policy issue. Income level, income distribution, unemployment, food self-sufficiency, and infrastructure are all different sides to the same problem. Second, developing countries are fairly heterogeneous in structures and interests. Structural differences have to be taken seriously in model building, otherwise results will be far from satisfactory as it has often happened.

Against this backdrop, the present contribution provides a cursory assessment view on policy analysis in agriculture in order to contribute to the debate in this area. It examines contemporary issues in agricultural policies, and the approaches for their analysis and assessment. Its intent is to trigger discussion rather than to capture all of the issues and details that deserve inquiry. The focus is mostly put on the international dimensions of agricultural policies and their impacts on economic development and agricultural markets. The ultimate goal is to improve our understanding in this relevant area for the development and welfare in all societies.

The reminder of the paper is structured as follows. In the next section, prominent policy issues setting the background for analytical modeling are described. These are mostly trade policy issues which triggered the need for and the development of modeling, and later extended to the analysis of other policy concerns related to agriculture, food and natural resources. The subsequent section depicts the facts surrounding multilateral trade, and the new regionalism as a trade policy outcome considered to be a direct consequence of the poor results obtained in multilateral trade. The following section discusses further relevant policy challenges related to agriculture, and the main applied economic models used to tackle them; obviously the different models are not presented in much detail since this would require various entire papers. The last section provides some final remarks.

Market Structures and Policy Measures

The need to consider structural differences between countries has to be taken seriously when evaluating the effects of policy measures using simulation tools. This is particularly true if the final welfare effects induced by the measures down to household level have to be evaluated. Such welfare changes follow the tracks of the existing structures, affecting the income and poverty level of all those involved in the process.

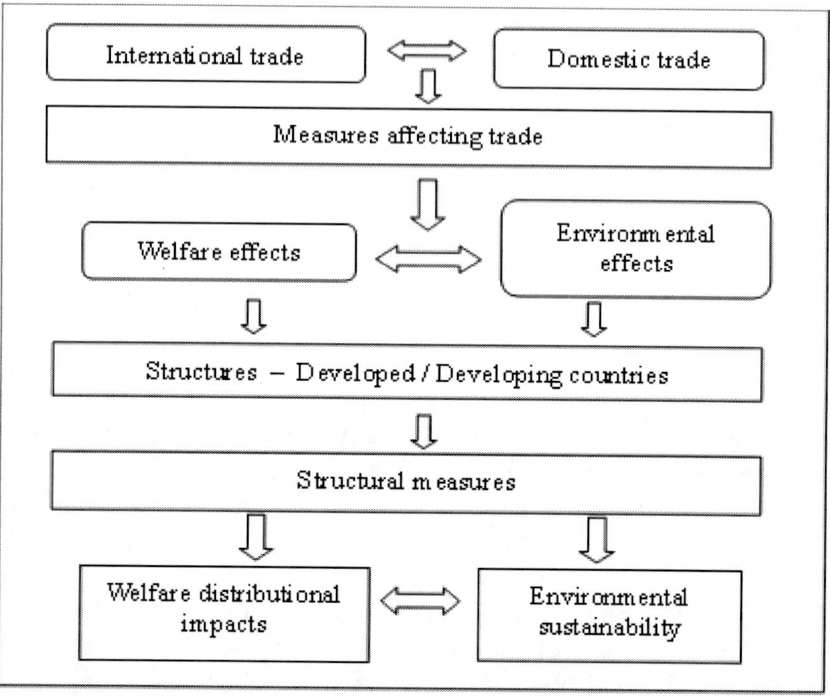

Figure 1. From trade to welfare and environment (The author).

Policy measures affecting domestic markets of major producing and importing countries and international commodity markets have important effects on prices and quantity of traded products. This results in potential welfare changes for producers and consumers. Obviously, welfare effects are not the same for different countries. It is not difficult to identify *prima facie* winners and losers following the rough role division mentioned above. But the problem does not stop there. Total positive welfare changes in the wake of policy changes, if they happen, have to be distributed not only among countries but also within countries. How are changes in international markets reflected in national economies? Which economic or social groups profit from such changes? Are poor farmers in developing countries better or worse off after the changes? The benefits and costs to different countries of using different agricultural policy instruments will depend on the structure of their individual economies as suggested in Figure 1.

In developed countries, where markets work rather fluently, institutional arrangements are transparent and transaction costs are relatively low, the usual measures of welfare gains or losses for different groups are roughly in line with predictable outcomes. Conversely, in developing countries with radically different economic and social structures, the effects might be quite different. Thus, in order to gauge the consequences on welfare of policy measures, other research instruments at household level need to be used. At household level, costs and benefits stemming from product and factor price shifts will translate into welfare changes depending on how households use their income. The choice of policy instrument requires the use of appropriate tools and analytical approaches, with the ultimate aim of drawing policy conclusions.

To detect how policy measures affect poverty, it is necessary to identify changes in welfare for different social groups. Such welfare distribution follows very different paths in developing and developed nations. Most models were originally developed for the latter group. If they are to be of any use for reaching policy goals regarding income, development and poverty alleviation, they must be carefully examined and modified so that welfare changes for target groups can be accurately estimated.

In developed countries, self-consumption of agricultural products by farmers is almost non-existent. Hence, the usual division between producers and consumers found in most models mirrors welfare distribution in a realistic way. In many developing countries this is not the case, though the self-consumption percentages vary widely between countries (Brooks et al. 2008). Something similar occurs with the supply of family labor. In competitive markets with homogeneous goods, it should make no substantial difference to the farm households whether labor is acquired in the market or provided by themselves. But when this is not the case due to high transaction costs or other constraints – as is often the case in developing countries, imperfect substitutability arises. This must be taken into account in policy modeling. Among market imperfections that make a difference to the situation in developed countries are access to credit, output, labor, and land markets.

It has been also documented that the standard treatment of farmers in industrial country models is not useful in developing countries, where a broader range of players should be considered within the farming sector. The farm household model is the tool that has frequently been used to adjust trade and policy modeling to the particular circumstances of developing countries. It has also been combined with general equilibrium models to draw the whole picture and identify the impacts of agricultural policies on different population sectors (Hertel and Reimer 2005).

Multilateral Trade Processes and New Regionalism

A comprehensive review of the methods and tools used to quantitatively evaluate the effects on production, trade and welfare of agricultural policies at national, regional and international levels shows the existence of a huge literature prompted particularly by multilateral trade negotiations in the 1980s and 1990s (Josling and Petit 2018). Since then, the need to consider policy alternatives and their possible effects created a huge demand for applied policy evaluation.

Subsequent analysis has shown that even with the inclusion of agricultural commodities under a more liberal trade framework, effective liberalization was not as great as expected. Border protection remained high and the tariff profile in many cases has become more complex Increased application of tariff-quotas has also introduced an unreliable system of market protection, acting sometimes in opposite directions. The structure of domestic protection has also shifted. It is highly concentrated in the EU, USA and Japan. Expenditure on allowed support has increased, in some cases resulting in trade distortions. Export subsidies have diminished but sometimes regulations have been fairly easy to circumvent.

The models originated in the wake of multilateral trade negotiations centered on the analysis of measures that directly affect trade and, more specifically, international trade. Other type of policy measures does not seem to have a direct effect on international exchanges, but they do affect markets and, consequently, prices and welfare of producers and consumers (Anania 2001). Set-asides, direct producer payments and other measures are among this group. Moreover, the concept of welfare as it is understood today goes beyond the generation of benefits from trade. For developing countries, once welfare is changed positively – as can be expected according to the liberalization theory, it still has to reach the sectors of society for which it is intended, especially the rural poor. This is a question that cannot be addressed with the tools adapted for developed nations.

More recently, the path of multilateral trade has been affected by commodity price increases and volatility, raising the question of how to reconsider the WTO framework within the new international price settings. Hence, will the multilateral/WTO rules, designed for a setting of cheap food, also be appropriate for a new one of high prices? Or is a fundamental refurbishing of international trade rules required? The new framework needs to be closely examined in order to reach agreements that will be long-lasting.

Parallel and partly consequence to these developments, there has been an intensification of the so-called new regionalism, and the internationalization of value chains. In fact, in the two last decades, probably the two most significant trends shaping the landscape of global trade, including agricultural trade, are the parallel development of the Regional Trade Agreements (RTAs), and the Global Value Chains (GVCs).

RTAs have proliferated rapidly after the impasse occurred in multilateral trade negotiations since the beginning of the 2000s. Although they are not a new phenomenon, they have become an important trade policy tool for virtually all countries. Both the number and the world share of trade covered under RTAs have been steadily growing over the last 20 years. Until January 2024 the WTO has received 594 notifications of RTAs, 361 of them were in force, up from 180 in 2005 (WTO 2024). The share of world trade developing within RTAs today exceeds 50% (OECD 2024a).

This trend is expected to be strengthened in the future by the numerous RTAs being proposed presently, and especially by the new mega-RTAs which have the potential to reshape much further the global trade landscape over the next years. Among the most significant of the latter are the Comprehensive and Progressive Agreement for Trans-Pacific Partnership (CPTPP) with partners in Asia, Australia, and the Western Hemisphere, and the African Continental Free Trade Agreement (AfCFTA) which is the world's largest regional trade agreement covering more than 1.3 billion people (Fusacchia et al. 2022, Ruta 2023). These mega-regional networks represent a move towards economic integration at a much greater scale than conventional RTAs, and a shift in the strategic focus of major trading powers from multilateral to regional deals. However, so far little reliable information is available about the content of these agreements, which represents a serious challenge for researchers when they attempt to perform *ex-ante* impact assessments for the different players involved and the products traded.

Admittedly, agriculture is traditionally one of the most difficult areas in which countries seek to achieve trade concessions. Agricultural trade regulatory frameworks affect the operating environment in which farmers, traders and consumers conduct their activities. They also have complex and far-reaching implications for broader public policy goals, including in the area of food security and rural development. Moreover, RTAs – and especially mega-RTAs, have direct implications not only within the member countries, but they are also likely to have consequences for third countries. Their systemic consequences can even be broader if their standards and norms become benchmarks for subsequent agreements, including those in the multilateral trading system. Hence it is important to consider RTAs not in isolation but instead in the broader context of the evolving global landscape of rules and policies affecting trade and investment in general.

Besides, while RTAs are likely to transform profoundly the established order of international trade, their actual impact on trade flows remains highly controversial. Existing literature show that likely impacts are highly variable (Disdier et al. 2014, Jean and Bureau 2015, Rodrik 2018). While some authors find that RTAs create large trade flows between signatories - though often at the expense of third parties (Tokarick 2008), others find that their impact tends to be much lower than frequently expected. This lack of robustness raises research questions and policy challenges about the actual consequences of a shift toward regionalism for global agricultural trade, i.e., there is a need for robust estimates of the shift to regionalism for a country that is considering adopting such a policy.

Among the main questions to answer since the beginning of this tendency have been whether the formation of regional trading-blocs raises or lowers welfare, and whether this development helps or hinders multilateral trade liberalization. Quantitative research in this area has focused on trade creation and trade diversion. Gravity models have been an important tool in applied research into the results of RTAs. The findings are not conclusive in supporting RTA benefits in terms of increased intra-bloc trade or trade creation in agricultural products.

Estimating the impact of these agreements also involves many practical issues, often owing to the lack of good-quality data and the methodological difficulties in correctly addressing so numerous interlinked variables, because frequently countries prioritize entering into agreements with partners with whom they have other reasons to wish strong commercial relations (e.g., geopolitical, cultural, historical ties). Currently, this is visible through the growing trends towards the *nearshoring* (Dewart 2022) and the *friendshoring* (Shibli 2024) in the international trade, where operators rely on fewer, trustful partners in response to wars and geopolitical conflicts. Indeed, in the present context of growing geopolitical tensions, calls for re-localization to favor shorter value chains – nearshoring – are increasingly frequent. Similarly, since the beginning of the war in Ukraine, there are trends promoting relocations to "friendly countries" – friendshoring – which are close politically, regulatorily and culturally, but not necessarily geographically.

In this regard, it is worth noting that the new international trade routes are getting longer due to increased geopolitical tensions, which in turn is increasing transport costs. Since November 2023, maritime traffic in the Red Sea has been disrupted by geopolitical tensions in the Middle East. As a result, some shipowners have announced significant price increases in this maritime zone through which 12% of international trade transits each year. Other operators have suspended their activities in the region, preferring to take the route to southern Africa, thus extending the route of vessels to destinations. Another focal point of world trade which is also congested is the Panama Canal where the passage of ships is slowed down significantly by the consequences of the drought which affected the country in 2023. In addition, the growing tensions in Asia have resulted in an increase in incidents in common maritime areas, with the subsequent search for safer routes. The lengthening of the routes of trade flows also increased in the wake of the rising of tensions between the two great powers China and the United States.

Other policy challenges relate to the balance between gains and losses induced by the proliferation of RTAs. For instance, while the promotion of free trade through preferential agreements can benefit economic development by integrating more communities into the world economy through reduced trade barriers and common standards, the construction of complex networks of non-MFN (Most Favored Nation) trade relations can increase discrimination and may erode transparency and predictability in international trade relations. It is therefore of systemic importance to shed light on this sort of dichotomy.

As for GVCs, it can be stated that their emergence has been one of the most visible results of the globalization and modernization of the world economy, and the agri-food sector is no exception. Presently, about 70% of international trade involves GVCs (OECD 2024b). Many companies divide their operations across the world creating international production chains that have altered the functioning of the production and trade patterns worldwide. Moreover, increased international product expansion, consumer concerns about health and food quality and the growing role of standards and market power of retailers, have brought stronger focus on the GVCs as an instrument to organize international product marketing schemes and institutions.

GVCs have changed many aspects of economic and trade relations while their policy ramifications are not fully addressed and understood (Franssen 2019). Changes reach not just the structure and composition of trade but also their implications for the role of the trade-disciplining and monitoring institutions. It should be noted that classical trade rules were designed when products were elaborated and exported fully or mostly by one country.

The rise of GVCs has promoted a rapid increase of trade flows in intermediate goods, which currently represent almost 75% of the imports of large economies like China. This reality in world trade has not been properly reflected in international trade statistics, which until recently had attributed the full commercial value of a product or service to the last exporting country, hence overstating the importance of the final producer in the value chain. Thus, more statistical information on how the world trade is performed (for instance the TiVA database by the WTO and OECD), together with the implications for individual countries is needed to understand these changes better, being this aspect a critical issue for researchers and policy-makers.

Empirically, there is a challenge to properly measure this ongoing segmentation of production process and its consequences and dependencies for trade flows (OECD 2023c). The existing literature (more theoretical than empirical) includes case studies for industrial products (very few detailed studies on agri-food products exist so far), and studies of trade in intermediate products based on national and more recently global input–output tables.

The debate on GVCs raises other issues including a better understanding of the motivations of firms, the use of economic policies, the critical importance of international logistics, and the influence of RTAs preferences on the patterns of trade developed by GVCs (Raimondi et al. 2023). Therefore, not only trade statistics but also trade policies must be consistently re-evaluated and updated to reflect the new structure of world trade and the operation of GVCs. Scrutiny should in addition contribute to close the gap between trade practices of businesses and the normative framework within which these practices take place.

Admittedly, the above-suggested issues on RTAs and GVCs need to take into consideration the changing global governance context, currently marked by the increasing pressure to integrate economic, environmental and social sustainability aspects of development, as stated, for instance, by the UN Sustainable Development Agenda which comprises 17 objectives many of whom relate to agriculture and the food system, including the role of trade policy in promoting gender equality (World Bank and WTO 2020). This also involves geopolitical conflicts such as the war in Ukraine and the Middle East. In this context, it should be recalled that though wealth created since the early 1990s by the globalized world economy is extensive, it came at a high cost, to a large extent due to the lesser attention paid to questions of equity and social inclusion, and an underestimation of asymmetries in capabilities among countries. Levels of inequality among and within numerous countries around the world are risky. Likewise, a high price has been paid in terms of insufficient consideration for the natural environment and climate change threats. As a corollary, linking global trade governance to sustainable development outcomes through solid metrics and indicators has become imperative.

Further Policy Challenges and Analytical Models

There are many other emerging challenges in agriculture that will require policy changes and further analytical efforts in the future. According to OECD (2023a) '*policies urgently need to be reformed to meet the triple-challenge of providing adequate, affordable, safe and nutritious food for a growing global population; providing livelihoods all along the food value chain; and doing so while increasing the environmental sustainability of the sector*'.

The challenges can be grouped within four main blocs: the problems posed by the increase in level and volatility of commodity prices; the growing importance of agriculture-based energy; environmental issues including global warming and sustainable use of resources; and the ever-present problem of trade reform and non-tariff barriers (Figure 2).

High commodity prices prompt the reflection that restrictive production policies and border protection put in place by the many developed countries are no longer useful. Flexibility seems to be a better response. It is important to consider to what extent higher world prices affect developing countries and how relevant they are for different socioeconomic groups, especially regarding the problems of sustainability and poverty alleviation (Mili and Arovuori 2023). Price spikes took place in an environment of increasingly open agricultural markets in some developing countries, exposing them to international price shocks. They hurt food importers, i.e., numerous developing countries, and some reactions such as recent export restrictions have destabilized agri-food markets and prevented food from being distributed to those in need (OECD 2024b). Moreover, production of biofuels putting pressure on demand for agricultural commodities shares the blame for the steep increase in agricultural prices. Increased production of ethanol and biodiesel has been driven more by policy incentives than economic fundamentals.

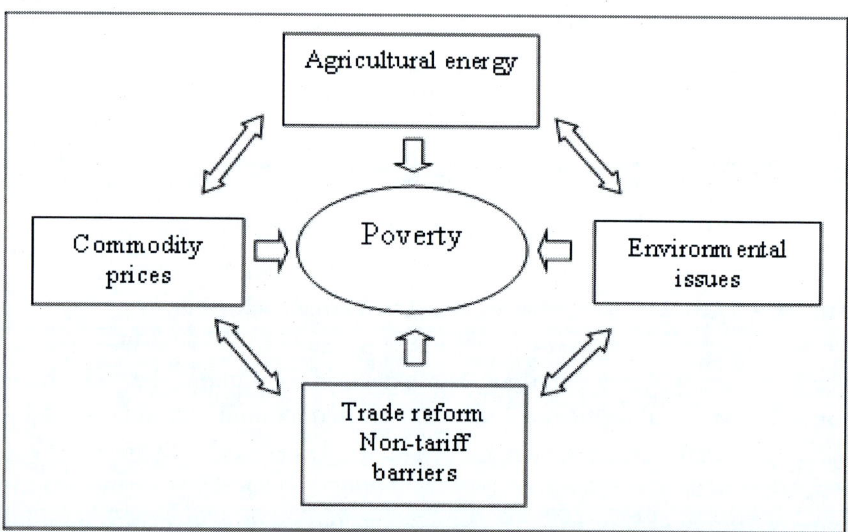

Figure 2. New challenges and agricultural poverty in international context (The author).

Policy simulation models made a major contribution to unravelling such complex relationships (Sumner et al. 2010). Both partial and general equilibrium models have been used for linking conventional agricultural production with agrofuel production. For instance, the LEITAP model is a general equilibrium, while ESIM, FAPRI, AGLINK/COSIMO are partial equilibrium agricultural models. RAUMIS and AGMEMOD, POLES and PRIMES were developed for the energy sector, and EUFASOM/ENFA for forestry. The mentioned partial equilibrium models that focus on agriculture do not include second generation biofuel crops, which are probably more suitable for obtaining energy than first generation ones since they are not in direct competition with food sources but compete for land and other inputs.

The LEITAP model is a global Computable General Equilibrium (CGE) model covering factor and product markets, often used to analyze WTO and CAP policy proposals. It is a modified version of the Global Trade Analysis Project (GTAP). The model has been extended to include the production, consumption and trade of biofuel products. It has been used to analyze the impact of EU policies regarding biofuels, such as the mandatory blending of ethanol with mineral fuels on world agricultural and fuel markets.

The POLES model is a partial equilibrium model that focuses on the energy sector and analyses greenhouse gas emitting activities. POLES is a dynamic model, developing a year-by-year recursive approach depicting pathways from 2005 to 2050. It yields long-term outlooks on energy demand, supply and prices, costs of abatement of CO_2 emissions and technology improvement scenarios. PRIMES is a modeling system for energy markets, with a similar but more detailed structure than POLES. It is particularly accurate in the description of technology choices and processing demand sectors, though it appears rather simple for the supply of energy crops.

The basic data needed to develop these models on biofuels are not readily available like the information on food commodities or other conventional raw materials, where agencies like the Statistics Division of FAO globally monitor product, demand and trade flows. Information is, however, available from private companies and institutions.

Besides, agricultural management practices and cropping patterns have a vast effect on nutrient cycles, freshwater availability and soil quality. Agriculture also plays an important role in emitting and storing greenhouse gases. From the modeling point of view, it is necessary to combine economic concepts and biophysical constraints in a consistent framework in order to quantify and analyze the long-term socio-economic and environmental consequences of different scenarios (Eickhout et al. 2008). CGE models are a valuable tool for simulating simultaneous equilibrium in a set of interdependent markets from a global perspective (Jafari et al. 2021a). CGE models show two different alternatives to achieve the goal of representing GHG emission and capture: a top-down CGE-model coupled with a bottom-up partial equilibrium agricultural land-use model that displays an improved functional structure within the CGE model itself. The main advantage of the integrated assessment (coupling) approach is the ability to benefit from the strength of partial equilibrium, which represents in detail agriculture and land use aspects in the economy-wide comprehensive framework of the CGE model. Integration poses major difficulties with data comparability, programming and inconsistencies. Internal extension of a CGE model through the introduction of new structural relations and corresponding parameters appears to be a more feasible and reliable method, though it is less accurate and realistic than the coupling approach.

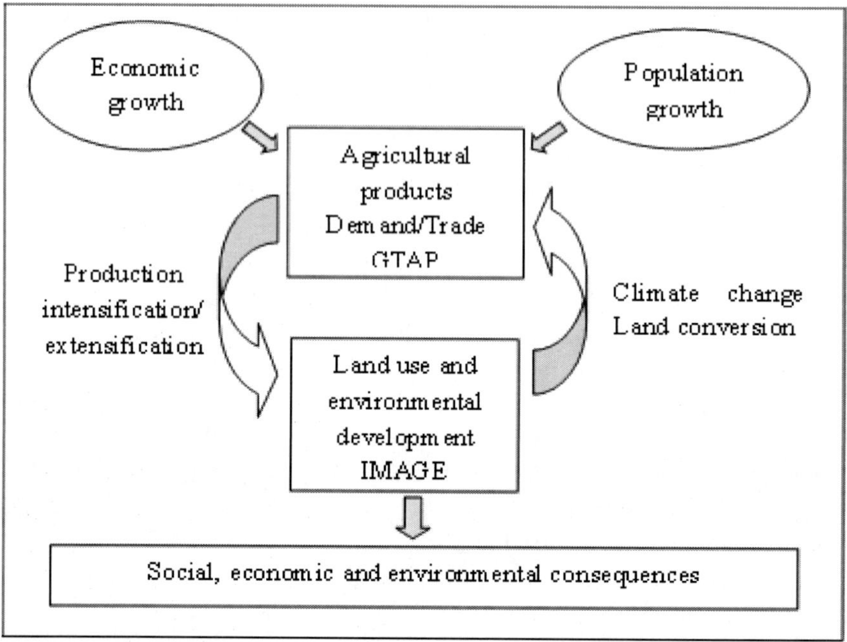

Figure 3. Agri-environmental modelling. Linking of LEITAP and IMAGE (Adapted from Eickhout et al. 2008).

The key challenge is to build a model framework that combines the advantages of a global approach – including the impact of non-agricultural sectors, and the full treatment of factor markets, including specific features of mobility and availability of land as a crucial factor in the GHG emission and capture representation. Eickhout et al. (2008) proposed such a solution coupling an expanded version of GTAP (LEITAP) with the Integrated Model to Assess the Global Environment (IMAGE), one of the most frequently used global land models to simulate land use emissions. IMAGE is a dynamic integrated assessment framework to model global change including an environmental perspective, quantifying the relative importance of major processes and interactions in the society-biosphere-climate nexus. Its three parts deal with energy consumption and fuels, linked to GHG emissions, with land use changes according to agricultural and forestry production (terrestrial system), and with the atmospheric-ocean system calculating changes in atmospheric composition considering oceanic CO_2 uptake. The linkage of both systems is depicted in Figure 3. The model has shown interesting effects on land supply under conditions of economic growth and its consequences for land rent, i.e., for the income situation of land owners and for countries where land is abundant or scarce. It also confirms the need to find, under certain circumstances, compromises between environmental conservation and immediate poverty reduction.

Moreover, it is worth highlighting that, methodologically, standard trade models often assume commodities to be homogeneous, and perfect substitution between products of different origin is the rule. Models designed under this assumption can only describe inter-industry trade, and countries are either importers or exporters of a certain good. Heterogeneous goods are distinguished by other factors than price and are seen as imperfect substitutes from the buyer's perspective. Hence, buyers are willing to pay different prices for different goods, and market partners can be buyers and sellers at the same time. The number of possible transactions doubles, at least in relation to models where goods are considered homogeneous, and much larger data and computing efforts are required. However, portraying the products as heterogeneous brings models much closer to reality, especially when the tasks to be modeled are bilateral in nature, such as trade disputes.

Product differentiation can be introduced according to product country of origin, following the Armington assumption widely used in CGE models. One problem arising from this is whether consumers really make distinctions amongst imported products. If so, this can only better reflect reality, but if it is only a modeling artefact, it introduces implicit distorting protection for domestic produce. Even if this heterogeneity of commodities on the basis of country origin reflects reality, an accurate estimation will depend on the accuracy of the matrix of cross-elasticities employed to quantify the effects. Alternative modeling tools to establish differences in apparently homogeneous produce have been developed more recently.

For agricultural products and commodities in general, perfect substitutability is an acceptable assumption and is recognized as such by a large range of models, among them AGLINK, FAPRI, SPEL-TRADE, SWOPSIM in almost all of its versions, WATSIM, and FAO's EFM. General equilibrium models use the product heterogeneity assumption, however. If the implicit protection for domestic production this assumption includes in the model is not real, a distortion is introduced. Therefore, it could be stated that consumers' preferences for products from certain origins is acceptable within a modeling exercise if it reflects a real observation, but not if it is a mere modeling instrument to force a more accurate result. Moreover, if a product is differentiated, this introduces a certain degree of market power in the structure of the model.

Conclusion

Policies play a crucial role in agricultural market structure and performance. In this contribution an attempt has been made to describing some stylized facts of the socioeconomic relevance of the prominent trends in agriculture from a policy analysis perspective. Some related questions for policy-making and research have also been explored in order to guide future reflection toward a better understanding of the evolving agricultural policy issues.

Arguably, sound agricultural policy analysis is steadily more necessary as policy issues related to agriculture and food have become increasingly controversial. Growing concerns have been expressed regarding the real impact of agricultural policy reforms, and whether the gains from policies exceed their costs. In this context, agricultural policy analysis provides timely information to relevant stakeholders and decision-makers.

In trade policy, researchers should provide policy-makers and other stakeholders with impartial, evidence-based analysis of the possible effects of GVCs, and both the already enforced RTAs and those under negotiations or ratification. Such *ex-ante* and *ex-post* analysis should contribute to the broader discussion on how rules and policies on agricultural trade could best promote food security and safety, rural development, social inclusion, and environmental sustainability, both in the countries that are seeking to become part of these networks and in those that are currently outside them.

Quantitative models are essential tools for representing the extremely complex interrelationships between agricultural systems, the environmental consequences of their operation and the economic results expected from them, within and across national and international borders. It is not usually possible to grasp such complexity with simple mental constructions. In addition, such biological, natural and socioeconomic structures react to policy intervention in ways that are difficult to predict. An established fact is that policy measures influence significantly welfare levels. To what extent and who is most or least affected by such changes is, however, open to debate. If environmental sustainability is also considered, the complexity of the analysis grows exponentially. Mathematical modeling of different types can contribute to policy discussions based on informed frames.

The strength of a number of models lies in their effectiveness in representing many agricultural policy instruments, even though the breadth of the models (products, markets) sometimes act against accuracy, mostly because of lack of good-quality data in many aspects. The partial equilibrium approach does not seem to be an impediment in these constructions applied to countries where the relative weight of the agricultural sector in the economy is not large. Partial equilibrium formulations have a pragmatic advantage over more complex tools (e.g., general equilibrium representations), allowing more effort to be dedicated to detailed analysis of specific measures.

It is also evident that the assumptions of perfect competition and fully rational decision-making appear to be more problematic. It is obvious that even commodities which are highly integrated in world markets like oilseeds and grains are not traded under fully competitive conditions. Moreover, the use of very simple behavioral equations for agents makes it necessary to consider representative agents, ignoring their diversity in developed countries' agriculture. In developing countries, this problem might lead to even larger distortions.

Econometric model results and their policy implications in developed countries hardly can be carried over to developing countries. This is mostly due to differences of agricultural structural conditions, the model parameters – especially elasticities – also might be substantially different, and the underlying assumptions – farms operating to maximize profit – might not apply in an environment where the farm household fulfills a dual role as producer and consumer. This leads to the development of other model structures like agricultural household models.

More specific criticism of modeling techniques points to the somehow excessively simplistic, exogenous way many models represent trade linkages. This implies that export or domestic prices in large trading blocs like the EU, the USA and China do not affect international prices, which is totally erroneous. A simulation using such a model would give largely distorted estimates of the effects of the policy analyzed. Moreover, it is a fact that much trade in agricultural commodities (bananas, grains) is practiced by very few large firms that are in a position to exert monopolistic market power. Few models take this element explicitly into account.

Another common feature of CGE models is the Armington assumption which is questioned on the ground that, assuming imperfect substitutability for the same commodity grown in different countries, a degree of trade protection for the product might be artificially introduced (Jafari et al. 2021b). Furthermore, rather than representing each policy explicitly, models are mostly represented synthetically by an equation which introduces a wedge between the international and the domestic price. This implicit modeling of policy instruments is generally considered unsatisfactory because of the number and variety of instruments involved.

Admittedly, models have numerous limitations, some are implicit to the tool used and others are the result of time and budget constraints that do not allow for suitable representations of the problem under scrutiny. Some conclusions have been drawn using for instance wrong tariff values - i.e., consolidated ones, when the tariffs actually applied were generally lower. When used for estimations, the premises, structures and particular cases have to be carefully considered. The most effective way to conduct simulations of agricultural policies would be to model the instruments explicitly and test them, one at a time, with adequate models such as partial economic spatial models using mathematical programming, or the general equilibrium models developed with this capability.

References

Anania G. 2001. *Modeling agricultural trade liberalization and its implication for the European Union.* Working paper 12. Istituto Nazionale di Economia Agraria.

Bastidas-Orrego L.M., Jaramillo N., Castillo-Grisales J.A., Ceballos Y.F. 2023. A systematic review of the evaluation of agricultural policies: Using prisma. *Heliyon* 9, e20292

Brooks J., Dyer G., Taylor E. 2008. *Modeling agricultural trade and policy impacts in LDCs.* OECD Food, Agriculture and Fisheries Working Papers 11. OECD, Paris.

Cardamone P. 2007. *A survey of the assessments of the effectiveness of preferential trade agreements using gravity models.* Working paper 07/09, Department of Economics and Statistics, University of Calabria.

Dewart J. 2022. *Nearshoring Offers Advantages During a Supply Chain Crisis.* Available at: https://www.supplychainbrain.com/blogs/1-think-tank/post/36086-nearshoring-offers-advantages-during-a-supply-chain-crisis

Disdier A.C., Fontagné L., Cadot O. 2014. North-South Standards Harmoni-zation and International Trade. *The World Bank Economic Review* 29(2): 327–352.

Eickhout B., van Meijl H., Tabeau A., Stehfest E. 2008. *The impact of environmental and climate constraints on global food supply.* GTAP Working Paper 47.

Franssen L. 2019. Global value chains and relative labour demand: A geometric synthesis of neoclassical trade models. *Journal of Economic Surveys* 33(4): 1232-1256.

Fusacchia I., Balié J., Salvatici L. 2022. The AfCFTA impact on agricultural and food trade: a value added perspective. *European Review of Agricultural Economics* 49(1): 237–284.

Hertel T., Reimer J. 2005. Predicting the poverty impacts of trade reform. *Journal of International Trade and Economic Development* 14: 377-405.

Jafari Y., Britz W., Guimbard H., Beckman J. 2021a. Properly capturing tariff rate quotas for trade policy analysis in computable general equilibrium models. *Economic Modelling* 104, 105620.

Jafari Y., Himics M., Britz W., Beckman J. 2021b. It is all in the details: A bilateral approach for modelling trade agreements at the tariff line. *Cand J Agr Econ.* 69: 415–442.

Jean S., Bureau J.C. 2015. *Do Regional Trade Agreements Really Boost Trade? Estimates for Agricultural Products.* CEPII Working Paper. Available at: http://www.cepii.fr/PDF_PUB/wp/2015/wp2015-09.pdf.

Josling T., Petit M. 2018. The Place of International Trade in Food and Farm Products in National Policy Discourse. In *Handbook of International Food and Agricultural Policies. Volume III: International Trade Rules for Food and Agricultural Products.* Edited by Karl Meilke and Tim Josling, 11-35. Singapur: World Scientific.

Mili S., Arovuori K. 2023. The Struggle to Sustain Agriculture in EU Southern Neighbourhood Partners: Improving Data and Policies. *EuroChoices* 22(1): pp. 50-57.

OECD. 2023a. *Agricultural Policy Monitoring and Evaluation 2023.* Adapting Agriculture to Climate Change. Available at: https://www.oecd-ilibrary.org/agriculture-and-food/agricultural-policy-monitoring-and-evaluation-2023_b14de474-en?utm_campaign=M%26E%20%2723 &utm_content=Read%20the%20report&utm_term=tad&utm_medium=email&utm_source=Adestra

OECD. 2023b. *Global value chain dependencies under the magnifying glass.* Available at: https://www.oecd-ilibrary.org/docserver/b2489065-en.pdf?expires=1710165775&id=id&accname=guest&checksum=83BFBCFD4CF6A5D3415ED2736453F981

OECD. 2024a. *Regional trade agreements are evolving – why does it matter?* Available at: https://www.oecd.org/trade/topics/regional-trade-agreements/

OECD. 2024b. *Global value chains and trade.* Available at: https://www.oecd.org/trade/topics/ global-value-chains-and-trade/

OECD. 2024c. *Agriculture and Trade Policy Research in 2023.* Available at: https://issuu.com/oecd.publishing/docs/oecd-agriculture-trade-policy-brief-21022024

Raimondi V., Piriu A., Swinnen J., Olper A. 2023. Impact of global value chains on tariffs and non-tariff measures in agriculture and food. *Food Policy* 118, 102469.

Rodrik R. 2018. What Do Trade Agreements Really Do? *Journal of Economic Perspectives* 32(2): 73–90.

Ruta M. 2023. The Rise of Discriminatory Regionalism. *Finance & Development* 60(2): 321-33.

Shibli A. 2024. *How will 'friendshoring' impact global trade in 2024?* Available at: https://www.thebanker.com/How-will-friendshoring-impact-global-trade-in-2024-1705912672.

Sumner D.A., Alston J.M., Glauber J.W. 2010. Evolution of the economics of agricultural policy. *American Journal of Agricultural Economics* 92(2): 403–423.

Tokarick S. 2008. Dispelling Some Misconceptions about Agricultural Trade Liberalization. *Journal of Economic Perspectives* 22(1): 199–216.

World Bank, World Trade Organization. 2020. *Women and Trade: The Role of Trade in Promoting Gender Equality.* Available at: Open Knowledge Repository (worldbank.org)

WTO. 2024. *Regional trade agreements.* Available at: https://www.wto.org/english/tratop_e/ region_e/region_e.htm

Index

E

F

G

Q

R

S

T

U

V

W

Y

Z